ENVIRONMENTALLY SIGNIFICANT CONSUMPTION

RESEARCH DIRECTIONS

Paul C. Stern, Thomas Dietz, Vernon W. Ruttan,
Robert H. Socolow, and James L. Sweeney, editors

Committee on the Human Dimensions of Global Change
Commission on Behavioral and Social Sciences and Education
National Research Council

NATIONAL ACADEMY PRESS
Washington, D.C. 1997

NATIONAL ACADEMY PRESS • 2101 Constitution Ave., N.W. • Washington, DC 20418

NOTICE: The project that is the subject of this report was approved by the Governing Board of the National Research Council, whose members are drawn from the councils of the National Academy of Sciences, the National Academy of Engineering, and the Institute of Medicine. The members of the committee responsible for the report were chosen for their special competences and with regard for appropriate balance.

This report has been reviewed by a group other than the authors according to procedures approved by a Report Review Committee consisting of members of the National Academy of Sciences, the National Academy of Engineering, and the Institute of Medicine.

This study was supported by Contract No. C R 823894-01 between the National Academy of Sciences and the U.S. Environmental Protection Agency, and Contract No. 50 DKNA 5-0005 between the National Academy of Sciences and the National Oceanographic and Atmospheric Administration. Any opinions, findings, conclusions, or recommendations expressed in this publication are those of the author(s) and do not necessarily reflect the view of the organizations or agencies that provided support for this project.

Library of Congress Cataloging-in-Publication Data

Environmentally significant consumption : research directions / Paul C. Stern ... [et al.],
    editors ; Committee on the Human Dimensions of Global Change, Commission on
    Behavioral and Social Sciences and Education, National Research Council.
        p. cm.
    Includes bibliographical references.
    ISBN 0-309-05598-9 (pbk.)
    1. Consumption (Economics)—Research. 2. Environmental policy—
Research. I. Stern, Paul C., 1944-   . II. National Research
Council (U.S.). Committee on the Human Dimensions of Global Change.
HB801.E66 1997
333.7—dc21   97-4847
                                                                                  CIP

Additional copies of this report are available for sale from the National Academy Press, 2101 Constitution Avenue, N.W., Box 285, Washington, D.C. 20055. Call 800-624-6242 or 202-334-3313 (in the Washington Metropolitan Area).

This report is also available on line at **http://www.nap.edu.**

Printed in the United States of America.

iii

The National Academy of Sciences is a private, nonprofit, self-perpetuating society of distinguished scholars engaged in scientific and engineering research, dedicated to the furtherance of science and technology and to their use for the general welfare. Upon the authority of the charter granted to it by the Congress in 1863, the Academy has a mandate that requires it to advise the federal government on scientific and technical matters. Dr. Bruce M. Alberts is president of the National Academy of Sciences.

The National Academy of Engineering was established in 1964, under the charter of the National Academy of Sciences, as a parallel organization of outstanding engineers. It is autonomous in its administration and in the selection of its members, sharing with the National Academy of Sciences the responsibility for advising the federal government. The National Academy of Engineering also sponsors engineering programs aimed at meeting national needs, encourages education and research, and recognizes the superior achievements of engineers. Dr. William A. Wulf is president of the National Academy of Engineering.

The Institute of Medicine was established in 1970 by the National Academy of Sciences to secure the services of eminent members of appropriate professions in the examination of policy matters pertaining to the health of the public. The Institute acts under the responsibility given to the National Academy of Sciences by its congressional charter to be an adviser to the federal government and, upon its own initiative, to identify issues of medical care, research, and education. Dr. Kenneth I. Shine is president of the Institute of Medicine.

The National Research Council was organized by the National Academy of Sciences in 1916 to associate the broad community of science and technology with the Academy's purposes of furthering knowledge and advising the federal government. Functioning in accordance with general policies determined by the Academy, the Council has become the principal operating agency of both the National Academy of Sciences and the National Academy of Engineering in providing services to the government, the public, and the scientific and engineering communities. The Council is administered jointly by both Academies and the Institute of Medicine. Dr. Bruce M. Alberts and Dr. William A. Wulf are chairman and vice chairman, respectively, of the National Research Council.

Contents

Preface

In scientific and political debates about the social causes of global change, blame for environmental degradation is often placed on the rapid growth in population numbers, particularly in the developing world. However, many social scientists have argued that per capita consumption in developed countries such as the United States is an equally serious cause of environmental degradation. The signing of international agreements to protect the atmosphere, such as the Framework Convention on Climate Change, has brought the analysis of consumption into the political arena. Developing countries have blamed northern consumption of fossil fuels for greenhouse gas emissions, and critics have pointed out that a child born in the United States will consume, on average, 10 times the resources and produce 10 times the pollution of a child born in Bangladesh or Bolivia.

Late in 1994, the U.S. Environmental Protection Agency, recognizing the environmental importance of consumption and the need for scientific study of the issue, asked the National Research Council for help in defining a research agenda on the global environmental impact of U.S. consumption. The National Research Council's Committee on the Human Dimensions of Global Change, which accepted the request, quickly realized that although a vast amount of potentially relevant research existed, the amount of empirical work focusing specifically on the environmental impacts of consumption and on the nature and causes of environmentally significant consumption was relatively small. We therefore decided that the most helpful approach would be to convene a small group of active researchers whose work would demonstrate a range of interesting and researchable scientific questions and who could help identify some promising directions for future research.

The committee held a workshop for this purpose in November 1995. The participants included researchers from government and universities. We were pleased that several international agencies and foundations expressed interest in the topic and sent representatives to the workshop. A number of scholars were asked to prepare draft papers that were circulated in advance and discussed at the workshop. It was clear from the lively interactions at the workshop that there are a variety of promising conceptual approaches to the study of environmentally significant consumption and that future research must address the complex interactions of factors such as technology, economics, culture, public policy, and individual behavior. This volume includes brief versions of some papers from the workshop and summarizes some of the ideas raised there about research strategies and directions. The papers show that the nature and level of resource consumption in the United States is causing environmental change and that consumption is influenced by a wide range of institutions, individual preferences, technologies, and economic policies. Yet there is no consensus on how and why environmentally significant consumption changes. Much could be learned from further quantitative analyses and qualitative case studies of specific consumption activities. Such research could illuminate the anthropogenic causes of environmental change and thus inform choices by government, the private sector, and individual citizens regarding ways to reduce the environmental impacts of consumption and to implement global environmental agreements. We hope that the publication of the results of this workshop will foster further discussion within the research and policy community and stimulate an expanded and fruitful research agenda on consumption as a factor in environmental and global change.

On behalf of the National Research Council Committee on the Human Dimensions of Global Environmental Change, I would like to thank Paul Stern, Thomas Dietz, Vernon Ruttan, Robert Socolow, and James Sweeney for organizing the workshop and editing this volume. I would also like to thank David Rejeski from the White House Office of Science and Technology Policy for suggesting such an interesting and challenging topic for a workshop. I also wish to thank Eugenia Grohman, Colene Walden, and Heather Schofield, all of whom provided essential help in editing and producing the volume, and Janine Bilyeu, without whose help with logistics and management the workshop would not have been so successful.

Diana Liverman, Chair
Committee on the Human Dimensions
of Global Change

1

Consumption as a Problem for Environmental Science

Paul C. Stern, Thomas Dietz, Vernon W. Ruttan,
Robert H. Socolow, and James L. Sweeney

For over two decades, the same frustrating exchange has been re-
peated countless times in international policy circles. A government offi-
cial or scientist from a wealthy country would make the following argument:

> The world is threatened with environmental disaster because of the de-
> pletion of natural resources (or climate change, or the loss of biodiver-
> sity), and it cannot continue for long to support its rapidly growing
> human population. To preserve the environment for future generations,
> we need to move quickly to control global population growth, and we
> must concentrate the effort on the world's poorer countries, where the
> vast majority of the population growth is occurring.

Government officials and scientists from low-income countries would
typically respond this way:

> If the world is facing environmental disaster, it is not the fault of the
> poor, who use few resources. The fault must lie with the world's
> wealthy countries, where people consume the great bulk of the world's
> natural resources and energy and cause the great bulk of its environ-
> mental degradation. We need to curtail overconsumption in the rich
> countries, which use far more than their fair share, both in order to
> preserve the environment and to allow the poorest people on earth to
> achieve an acceptable standard of living.

Both parties to this stylized debate agree about the importance of
acting to reverse environmental degradation in the world and of finding

development paths that preserve environmental quality. They also agree that knowing what action to take requires an understanding of the causes of environmental degradation. They disagree, of course, on what is the correct understanding.

The disagreement is noteworthy in two respects: the "either-or" framing of the problem (either environmental threats result from overpopulation by the poor or overconsumption by the rich), and the limited role played in the discussions by scientific analysis of human-environment interactions. In contrast to the major international efforts that have been mounted to understand the biogeochemical processes that account for global climate change, acid deposition, ozone depletion, loss of biodiversity, and other environmental threats, relatively little scientific attention has been given to understanding the economic, social, cultural, and institutional processes that set such anthropogenic environmental changes in motion. It is well known, for instance, that world population growth is concentrated in the poorer countries and that consumption of key resources such as oil is concentrated in the richer countries. Too often, however, analysis of the environmental impacts of human activity has not gone much farther than this, and the policy debate has suffered as a result. Governments and citizens around the world need far more detailed knowledge to effectively anticipate and cope with environmental threats. Specifically, they need empirical analysis in two areas: the particular human choices and actions most responsible for adverse changes in the biophysical environment and the potential for addressing the threats by affecting those choices and actions. This is not to say that knowledge alone is sufficient for effective coping. Knowledge is necessary, but it is also necessary to reconcile interpretations of knowledge, to forge agreements on how to act in spite of uncertainty and conflicting values and interests, to command resources, and so forth.

Scientists have long recognized that the magnitude of anthropogenic environmental changes depends on both human population size and on what this population does. This relationship was conceptualized in the early 1970s as follows:

$$I = P \times A \times T,$$

in which I represents environmental impact, P represents population, A—the economic output per capita—is usually interpreted as a measure of affluence, and T—the environmental impact per unit of economic output—is sometimes interpreted as a characteristic of technology (Ehrlich and Holdren, 1971; Commoner, 1972; Holdren and Ehrlich, 1974).

As stated, this equation is a tautology: environmental impact equals population times (economic output per unit population) times (environmental impact per unit economic output). However, the equation does suggest that one should consider at least three factors in analyzing anthropogenic environmental changes: the population, some measure of activity per person, and some measure of the average impact of each unit of activity on the environment. This so-called IPAT identity has been used as a basis for analysis, but it must be treated with care. While it must hold mathematically, one cannot simply use the equation for prediction because these three factors do not develop independently of one another. For example, as per capita economic output increases in a society, that society may strive to reduce environmental impacts and succeed in decreasing the environmental impact per unit of economic activity. Similarly, increasing population may alter the economic output per capita. The interdependencies of P, A, and T can seriously limit the usefulness of the equation, especially for analyses over long time periods. The IPAT formulation has also been controversial because of the way its terms have been interpreted. For instance, the interpretation of T as "technology" has been criticized for drawing attention away from the roles of social and economic institutions in environmental degradation. For a recent discussion of these issues, see Dietz and Rosa, 1994.

With these caveats in mind, treating environmentally relevant human activity as a mathematical product is intuitively appealing because it reduces the analytic problem to two factors: numbers of people (population), and the impact of the average person on the environment (sometimes referred to as the impact of consumption). Population is by far the easier of the two concepts to study. Its units of measurement are obvious and there is an established scientific discipline, demography, that studies human population dynamics. Consumption, by contrast, has neither well-defined and accepted units of measurement nor a scientific community devoted to studying its dynamics. As noted in Chapter 2, it even lacks a shared definition that is useful for studying environmental effects.

This book results from a workshop that brought together specialists from a variety of fields to discuss research on environmentally significant "consumption" and its causes, with special focus on the United States and other wealthy countries. Participants discussed research that aimed to improve knowledge of consumption phenomena and thus to strengthen the knowledge basis for policy planning. This volume, like the workshop, attempts only a small step toward useful knowledge. It does not attempt to offer practical conclusions about how to alter consumption patterns or to definitively map the intellectual domain. In fact, as noted below, there are several active and relevant areas of research that are not discussed here. The volume does, however, demonstrate that environmentally significant consumption requires careful scientific analysis, and it tries to convey to the reader some of the excitement of taking a scientific approach to the issue by noting the potential of some relatively untraveled but promising paths to new understanding. It points out some ways to pursue these research directions, and we hope it will stimulate action by researchers and research sponsors.

FOUR CLASSES OF EMPIRICAL QUESTIONS

The broad topic, "what each person does to the environment," divides fairly readily into four major classes of empirical questions, each of them researchable and each critical for understanding and controlling the environmental impacts of consumption.

1. Which human activities are the significant environmental disruptors? How environmentally significant is each activity, and in what ways is it destructive? What have been the trends of these activities over time, and how may technological change and other forces affect those trends in the future?

Some human activities are well known to be environmentally disruptive. Fossil energy consumption, the most obvious example, is a major contributor to global climate change, urban air pollution, and acid deposition. Mining and processing of metals, especially toxic heavy metals, pollutes water and threatens human health and ecological systems. Agricultural practices such as fertilization, irrigation, and the use of pesticides pollute water and alter the nitrogen and fresh water cycles. Many other human activities also have environmental significance. The first questions about the environmental impacts of human activities concern which of these activities are important enough in terms of their environmental consequences to deserve serious and immediate attention and what kinds of impact each has.

These questions have great practical importance. If the concern is to limit certain destructive environmental impacts of what people do, decision makers need to know enough to set priorities among possible target activities, to understand the tradeoffs involved in targeting one set of activities rather than another, and to consider the potential environmental impacts of the activities that might replace those the policy makers seek to control. These decision makers need to know which human activities are significantly disturbing environmental systems and which activities are likely to do so in the future. To respond to such conditions and possibilities, they need more detailed knowledge. For example, decision makers who want to protect an aquatic ecosystem need to know whether it would be more effective to reduce fertilization so that less nitrogen leaches off nearby farmland or to promote better soil conservation practices to control the amount and timing of runoff. Those who want to reduce the environmental impact of the automobile industry need to consider the tradeoffs involved in building vehicles of less steel and more plastic. Lighter cars use less fuel and require less mining and smelting, but the life cycle of plastics has its own, possibly offsetting, environmental effects. Decision makers also need to consider the secondary effects of possible policies. A decision to make all new cars lighter might induce some consumers to purchase light trucks instead of cars, thus undermining the intended result. Addressing these questions requires, among other kinds of understanding, knowledge of the environmental effects of each kind of human activity that policy may affect. There are considerable bodies of relevant research on the environmental consequences of particular human activities, on natural resource accounting, and on the incorporation of environmental factors into national income accounts. These issues are not, however, the focus of this volume.

2. Who are the key actors responsible for the environmentally disruptive activities? Which of their actions are the important ones?

To limit the environmental impact of any human activity, it is essential to understand which types of individuals or organizations account for most of the activity and how the activity of interest fits into their overall purposes.

A single environmental problem may result from different actors doing different kinds of things. For example, urban air pollution consists partly of ozone, much of which is a by-product of emissions from motor vehicles operated by individuals. But air pollution also consists partly of sulfur oxides, for which individual action bears little responsibility. Sulfur oxides come largely from coal combustion, which in the United States is mainly an activity of large industrial organizations, especially electric utility companies. Thus, controlling air pollution may require the implementation of very different policies, each suited to a particular class of actor and kind of activity.

It is also important to know which of the things an actor does are environmentally important. For instance, if one wants to consider options for reducing water consumption among residences in a municipality, it is useful to know how much of that use is for watering lawns, filling pools, washing cars, bathing, washing, cooking, and other purposes, because households may treat some of these uses as more essential than others or find some of them relatively easier to postpone. If a large proportion of water is used to do things that can be easily postponed, a time restriction may be an effective intervention; if a large proportion is for uses considered nonessential, a price incentive may be highly effective.

3. What forces cause or explain environmentally disruptive actions?

Human activities that alter the environment respond to a mixture of social, economic, technological, political, and psychological forces. The example of energy use in residences illustrates the general situation. Energy use depends on multiple factors within households, including the number of people in the household and on whether any of them spend their days at home, which affect the demand for heating, cooling, and the services that appliances provide. It depends on household income, which affects the size of a dwelling and therefore its energy demand, and also affects the household's ability to invest in energy-efficient home technology, which can have a countervailing effect. The use of energy also depends on the age and sex of the occupants, which affect the desired ambient temperature. And it depends on household members' desires for appliances; their attitudes, beliefs, and values concerning energy use, frugality, and various other matters; and even their cultural backgrounds. In addition, energy use depends on household technology and its relation to the physical environment: the appliances being used and their designs, the home's construction, its exposure to wind and weather, and its surrounding micro- and macroclimates.

Household energy use is influenced by many additional factors as well, which create the context for key choices and actions in the household. An obvious one is the price of energy, which is affected, in turn, by public policies of energy taxation and utility regulation, the competitiveness of energy industries, advances in the technology of energy production and distribution, and perhaps the history of national energy production—in the United States a long history of energy self-sufficiency may help explain the strength of the political forces that have for two decades stymied efforts to raise oil prices to help meet environmental objectives. Energy use is also affected in indirect but important ways by the standard practices of the home construction and appliance manufacturing industries, by local building codes, and by the practices of home mortgage lenders, which may or may not offer financial benefits to the buyers of energy-efficient homes that cost less to maintain. Consumption is affected by tax incentives for home ownership because incentives make larger homes more affordable, and these homes use more energy. And energy use may also be affected by policies intended to influence it directly, such as regulations governing appliance manufacture and the information and financial incentives that governments and energy suppliers have sometimes offered to households to induce them to invest in energy efficiency.

This long but incomplete list suggests several things: that an environmentally significant consumption activity like household energy use is multiply determined; that the influences are of many kinds, both direct and indirect; that the many influences are interdependent, acting in combinations rather than additively; that it will take many disciplines working together to understand how they drive the phenomenon; and that the influences act on different time scales, with some, like the demand for heating and cooling, capable of changing in minutes, hours, and days, while others, like those affecting building construction, have effects that last for decades. In these respects, energy use is much like other environmentally relevant human activities and choices (National Research Council, 1992: Chapter 3). Each of these choices and activities responds to multiple influences, yet for each, the causes are amenable to scientific study.

4. How can environmentally disruptive human activities be changed?

From a policy perspective, this question provides a sufficient motive for asking all the previous ones: if decision makers must contemplate changing environmentally disruptive activities, they need to understand the nature and causes of those activities. Knowing how to change environmentally significant activity, however, requires more than an understanding of the causes. Those who would be influence agents need enough knowledge about how these activities might be changed to select effective options. Some knowledge already exists about the effects of particular kinds of intervention from studies of the diffusion of technological innovation (for reviews of this literature see, e.g., Rogers, 1995; Ruttan, 1996); the effects of environmental regulation, monetary inducements and penalties, and other incentive-based interventions (e.g., Baumol and Oates, 1988; Cropper and Oates, 1992; Geller et al., 1982; Nichols, 1984; Tietenberg, 1985); the effectiveness of information in promoting proenvironmental behavior change (e.g., Katzev and Johnson, 1987; Gardner and Stern, 1996); the role of social movements in environmental change (e.g., Brulle, 1996; Dunlap and Mertig, 1992); and institutional strategies for environmental management (e.g., Ostrom, 1990; North, 1994). Much more work needs to be done, however, to develop these insights further, to compare the effectiveness of different types of interventions against each other, and to understand the potential synergisms that may arise from combining different types within a single, coordinated intervention (e.g., Gardner and Stern, 1996:Chapter 7).

PURPOSES AND STRUCTURE OF THE BOOK

This book, though inspired by policy makers' concerns about how to reduce "the environmental impacts of U.S. consumption," focuses on building basic knowledge rather than directly on policy questions. Our concern is with understanding which human activities are of major environmental significance, what forces shape those activities, and what their trends have been and might become in the absence of policy intervention. This sort of understanding is essential for policy makers because it can help them define the important policy issues, anticipate needs for intervention, and identify appropriate targets for interventions. It is a large task to build such understanding, and our modest purpose here is to encourage progress in that direction. Consequently, we do not advocate policy strategies for controlling the environmental impacts of consumption or discuss the extensive literature on the effectiveness or social acceptability of policies that have been proposed or enacted to accomplish that end.

Much work has already been done on some issues important to understanding environmentally significant consumption, and we do not attempt to summarize it here. The relevant areas that are already well developed or currently active include work on the effects of prices and other economic signals (e.g., Baumol and Oates, 1988; Cropper and Oates, 1992; Nichols, 1984), on trends toward "dematerialization" and "decarbonization" in the economy (e.g., Herman et al., 1989; Nakićenović, 1996; Wernick et al., 1996), on the potential for shifting materials flows from a linear pattern to a cyclic one that uses wastes as inputs to production (e.g., Allenby and Richards, 1994; Frosch, 1996; Socolow et al., 1994), on the environmental impacts of international trade (e.g., Runge, 1995; Runge et al., 1994), and on indicators that might allow systematic comparison of the environmental impacts of different kinds of human activity (e.g., Fava, 1991; National Research Council, 1994; Lave et al., 1995; Wackernagel and Rees, 1996). This book touches on some of these areas but focuses primarily on two other things: it tries to suggest the outlines of a scientific field, "environmental impacts of consumption," that may someday bring together these areas of work as well as additional areas within a coherent intellectual framework; and it presents illustrations of some of the other kinds of research, not yet well developed, that may eventually make important contributions to that field.

The book is intended for people who want to improve understanding of the human activities that constitute environmentally significant consumption: scientists in fields that can build this understanding and the organizations that might support their scientific work. It shows that, in addition to the knowledge that is already being developed, there are many critical questions that have barely begun to be addressed and, accordingly, that there are major opportunities to build useful knowledge. We hope it captures some of the excitement of the workshop from which it came, in which people from many separate intellectual fields began to see their common interests, and that it encourages the exchange of ideas necessary to build the needed understanding.

Scientific endeavors normally begin by adopting a working definition of the phenomenon to be examined, but the field of consumption and the environment has not yet made this step. As Paul Stern shows in Chapter 2, consumption is an ambiguous concept. It has precise and distinct meanings in physics, economics, and ecology and somewhat less precise meanings in sociology, but none of these corresponds to the usage in the phrase "environmental impacts of consumption." Stern discusses two definitions of consumption that might guide research on its environmental impacts and offers one of them as preferable. His analysis suggests the breadth of the field by noting a variety of open research questions that require attention from the social and natural sciences.

Chapters 3 and 4 present brief reports, taken from presentations at the workshop, that suggest some interesting research directions. The reports in Chapter 3 address issues of measuring and tracking flows of materials and energy that are affected by human consumption activities; those in Chapter 4 concern the driving forces of environmentally significant consumption. The reports indicate some promising directions for research; in addition, each one includes numerous citations that can direct an interested reader farther into the particular domain of study. We see these reports as suggestive. They do not present a full menu of research opportunities but only a list of hors d'oeuvres. We hope these reports whet the appetites of researchers and research sponsors.

Chapter 5 discusses how to set research priorities. It proposes an importance criterion for agenda setting: that top priority go to research on aspects of consumption with major environmental effects. This criterion suggests the strategy of identifying the most environmentally disruptive human activities and then searching to explain them and to account for how they affect the environment. The chapter examines this strategy in some detail, noting how some of the reports in Chapters 3 and 4 exemplify its use. The chapter also discusses two other ways to identify important research topics. One, which focuses on possible policy interventions, is more useful for policy analysis than for basic understanding of consumption. The other begins with social phenomena and works through to their environmental implications. Although this last strategy can yield useful insights that might not come from a research program that starts with environmental changes, it places a burden of proof on the researcher to demonstrate that the environmental effects are important and not just plausible or theoretical.

REFERENCES

Allenby, B., and D.J. Richards, eds.
  1994   *The Greening of Industrial Ecosystems.* National Academy of Engineering. Washington, D.C.: National Academy Press.
Baumol, W.J., and W.E. Oates
  1988   *The Theory of Environmental Policy,* 2nd ed. Cambridge, England: Cambridge University Press.
Brulle, R.J.
  1996   Environmental discourse and social movement organizations: A historical and rhetorical perspective on the development of U.S. environmental organizations. *Sociological Inquiry* 66:58-83.
Commoner, B.
  1972   The environmental cost of economic growth. Pp. 339-363 in *Population, Resources, and the Environment.* Washington, D.C.: U.S. Government Printing Office.
Cropper, M.I., and W.E. Oates
  1992   Environmental economics: A survey. *Journal of Economic Literature* 30:675-704.
Dietz, T., and E.A. Rosa
  1994   Rethinking the environmental impacts of population, affluence, and technology. *Human Ecology Review* 1:277-300.
Dunlap, R.E., and A.G. Mertig
  1992   *The U.S. Environmental Movement, 1970-1990.* Washington, D.C.: Taylor and Francis.
Ehrlich, P.R., and J.P. Holdren
  1971   Impact of population growth. *Science* 171:1212-1217.
Fava, J.A., R. Denison, B. Jones, M. Curran, B. Vigon, S. Selke, and J. Barnum, eds.
  1991   *A Technical Framework for Life-Cycle Assessments.* Washington, D.C.: Society for Environmental Toxicology and Chemistry.
Frosch, R.A.
  1996   Toward the end of waste: Reflections on a new ecology of industry. *Daedalus* 125(3):199-212.
Gardner, G.T., and P.C. Stern
  1996   *Environmental Problems and Human Behavior.* Boston: Allyn and Bacon.
Geller, E.S., R.A. Winett, and P.B. Everett
  1982   *Preserving the Environment: New Strategies for Behavior Change.* New York: Pergamon.
Herman, R., S.A. Ardekani, and J.A. Ausubel
  1989   Dematerialization. Pp. 50-69 in J.H. Ausubel and H.E. Sladovich, eds., *Technology and Environment.* Advisory Committee on Technology and Society, National Academy of Engineering. Washington D.C.: National Academy Press.
Holdren, J.P., and P.R. Ehrlich
  1974   Human population and the global environment. *American Scientist* 62:282-292.
Katzev, R.D., and T.R. Johnson
  1987   *Promoting Energy Conservation: An Analysis of Behavioral Research.* Boulder, Colo.: Westview.
Lave, L.B., E. Cobas-Flores, C.T. Hendrickson, and F.C. McMichael
  1995   Using input-output analysis to estimate economy-wide discharges. *Environmental Science and Technology* 29:420A-426A.
Nakićenović, N.
  1996   Freeing energy from carbon. *Daedalus* 125(3):95-112.

National Research Council
   1992   *Global Environmental Change: Understanding the Human Dimensions.* P.C. Stern, O.R. Young, and D. Druckman, eds. Committee on the Human Dimensions of Global Change. Washington, D.C.: National Academy Press.
   1994   *Assigning Economic Value to Natural Resources.* Papers presented at the Workshop on Valuing Natural Capital for Sustainable Development, July 1993. Washington, D.C.: National Academy Press.
Nichols, A.L.
   1984   *Targeting Economic Incentives for Environmental Protection.* Cambridge, Mass.: MIT Press.
North, D.C.
   1994   Constraints on institutional innovation: Transaction costs, incentive compatibility, and historical considerations. Pp. 48-70 in V.W. Ruttan, ed., *Agriculture Environment and Health: Sustainable Development in the 21st Century.* Minneapolis: University of Minnesota Press.
Ostrom, E.
   1990   *Governing the Commons: The Evolution of Institutions for Collective Action.* Cambridge, England: Cambridge University Press.
Rogers, E.M.
   1995   *Diffusion of Innovations,* 4th ed. New York: Free Press.
Runge, C.F.
   1995   Trade, pollution, and environmental protection. Pp. 353-375 in D.W. Bromley, ed., *The Handbook of Environmental Economics.* Oxford, England: Blackwell.
Runge, C.F., F. Ortalo-Magné, and P. Vande Kamp
   1994   *Free Trade, Protected Environment: Balancing Trade Liberalization and Environmental Interests.* New York: Council on Foreign Relations.
Ruttan, V.W.
   1996   What happened to technology adoption-diffusion research? *Sociologia Ruralis* 36:1-24.
Socolow, R., C. Andrews, F. Berkout, and V. Thomas, eds.
   1994   *Industrial Ecology and Global Change.* Cambridge, England: Cambridge University Press.
Tietenberg, T.H.
   1985   *Emissions Trading: An Exercise in Reforming Pollution Policy.* Washington, D.C.: Resources for the Future.
Wackernagel, M., and W. Rees
   1996   *Our Ecological Footprint.* Gabriola Island, B.C., Canada: New Society Publishers.
Wernick, I.K., R. Herman, S. Govind, and J.H. Ausubel
   1996   Materialization and dematerialization: Measures and trends. *Daedalus* 125(3):171-198.

2

Toward a Working Definition of Consumption for Environmental Research and Policy

Paul C. Stern

The concept of "environmental impacts of consumption" is rooted partly in environmental high politics. These roots can be discerned in a 1994 Presidential Decision Directive that first mobilized the U.S. government to pay attention to consumption as an environmental issue. The directive was issued in preparation for the International Conference on Population and Development that would be held in Cairo that October, at which it was widely expected that any U.S. initiative on controlling population growth would be met by criticism directed at American levels of consumption. The directive stated that the United States and other developed countries must maintain an awareness of their disproportionate impacts on the global environment through consumption patterns that are at several times the level of developing countries. To effectively achieve the goal of marshalling an international response to population growth trends, it said that the United States must also demonstrate leadership by example in addressing the implications of these consumption patterns, with an aim toward reducing the negative global environmental impacts of consumption of goods and services in the United States. The U.S. Environmental Protection Agency (E.P.A.), in coordination with the Departments of Energy and Transportation and other appropriate agencies, was directed to develop a statement articulating U.S. strategies for reducing such negative impacts. The directive went on to give the E.P.A. responsibility for developing a research agenda to guide future policy in this area.

In this political usage, "environmental impacts of consumption" appears to refer to everything people do, aside from increasing their numbers, that may harm the environment. This usage especially emphasizes what people in rich countries do. Treating the subject scientifically, however, requires a more precise definition of "consumption" that is acceptable across disciplines and is useful for analyzing the environmental impacts of human choices and actions. I discuss some specialized disciplinary meanings of consumption, the inadequate definition implicit in much recent discourse on the subject, and finally a working definition that I tentatively propose for use in environmental research and policy.

SPECIALISTS' MEANINGS OF CONSUMPTION

Consumption has fairly precise meanings in several scientific communities that are likely to address the "global environmental impacts of consumption." These meanings are in common use in their respective disciplines, whose adherents often have them in mind when discussing the environmental impacts of consumption. Unfortunately, none of these disciplinary meanings corresponds to the one in the phrase. A good way to begin to clarify thinking is to state these definitions, because they differ from what "consumption" in the quoted phrase seems to mean.

The Physicists' Meaning

According to the First Law of Thermodynamics, consumption is impossible: Matter/energy can be neither produced nor consumed. So for physicists, consumption must be translated as *transformations* of matter/energy. According to the Second Law, such transformations increase entropy, and this increase in entropy, to the extent that it takes the form of pollution or of a decrease in the usefulness of the transformed resource, is part of what is meant by "environmental impacts of consumption."

The Economists' Meaning

Economists define consumption as part of total economic activity: it is total spending on consumer goods and services (e.g., Samuelson and Nordhaus, 1989:969). The rest of economic activity consists of investment in capital goods. Economists also distinguish the consumption of goods and services from their production and distribution. In neither of these senses does the economists' usage conform to what is meant in the phrase "environmental impacts of consumption." Investment has environmental impacts just the way the purchase of consumer goods and services does. In fact, the activities may be physically identical, as when a truck or a computer is produced and purchased—either for use as capital equipment in a firm or as a consumer good. And in economists' terms, the environmental impacts of "consumption" result from production and distribution as well as from economic consumption. All three processes have environmentally significant impacts, and production and distribution may be more environmentally disruptive than consumption. Certainly, production processes such as mining and agricultural tillage are responsible for significant pollution problems, and also significantly degrade natural resources (in more precise economic terms, they make resources increasingly costly to transform for productive purposes). The economists' definition of consumption leads many economists to consider it analytically inappropriate to speak of "environmental impacts of consumption" because this phrasing artificially extracts consumption from the system of activities of which it is a part. These economists might prefer to translate the environmental impact of "consumption" as the environmental impact of *economic activity*. This formulation reflects the systemic unity of economies and also suggests that there may be differential impacts of different kinds of economic activity. For example, the impact of the average dollar invested may be different from that of the average dollar spent on consumer goods and services; different investments may have different impacts; spending on goods may have a different impact from spending on services, and so forth.

Confusion sometimes arises when people use economic statistics, which apply the economic definition of consumption, to analyze the environmental effects of "consumption." They may, for example, treat an increase in consumer spending as if it automatically indicates a proportional increase in environmental impact. This may or may not be so, depending on what changes in the size and types of economic consumption account for the increased spending.

The Ecologists' Meaning

To ecologists, green plants are (primary) producers, and humans and other animals are consumers. (Humans also "consume" minerals.) Ecologists define production, or net primary productivity (NPP), in terms of photosynthesis: "NPP is the amount of energy left after subtracting the respiration of primary producers (mostly plants) from the total amount of energy (mostly solar) that is fixed biologically" (Vitousek et al., 1986:368). In this meaning, any organism that obtains its energy by eating is a consumer; human consumption corresponds to what humanity does with the estimated 40 percent of global terrestrial NPP that we "appropriate" (Vitousek et al., 1986). It is not obvious, however, that the 40 percent estimate is a valid measure of the global environmental impact of human consumption, because human appropriation of primary productivity is not simply an ecological negative. Humanity transforms ecosystems, substituting species that seem to meet our needs for those that do not. In the process, some species become more prevalent, and in some cases, productivity increases. For example, agriculture adds nutrients to the soil and provides additional habitat for alfalfa weevils, honeybees, aphids, and the like, and for their predators and diseases. So the link between human consumption of global NPP and the "environmental impacts of consumption" is not 1:1. The two usages are not equivalent, and their relationship is yet to be determined.

Meanings in Sociology

"Consumption" also has sociological meanings, not precisely defined, that are reflected in terms like "consumerism" and "conspicuous consumption." In this usage, "consumption" connotes what individuals and households do when they use their incomes to increase social status through certain kinds of purchases (see, e.g., Veblen, 1899; Campbell, 1987; Scitovsky, 1992; Institute for Philosophy and Public Policy, 1995). Consumption in this sense is not related in any straightforward way to environmental impact, as can be seen by looking at what may be included in conspicuous consumption. In some American subcultures, one can increase status by building an all-solar house that conspicuously consumes money (for architectural design, solar panels, and so forth) but that may reduce environmental impact if it decreases fossil and nuclear energy consumption enough to compensate for the additional materials in the house. Similarly, a late-model luxury car may cost more money, provide more status, and yet consume less fuel and steel than an old pickup truck. The sociological definitions have different referents from those implied in the phrase "environmental impacts of consumption" because they do not distinguish environmentally benign from environmentally destructive consumption. It is analytically misleading to presume that manifestations of consumerism are necessarily destructive to the environment. The recent phenomenon of "green consumerism," which encompasses choices that are, or are believed to be, environmentally beneficial, illustrates the point.

A POPULAR BUT INADEQUATE DEFINITION

As a step toward a working definition of consumption, it may help to explicate a definition that is implicit in much popular discussion of consumption and the environment, that implies an interesting research agenda, but that ultimately provides an incomplete and misleading basis for analyzing the issue. I do not advocate accepting this definition.

Rather, I present it because it embodies some of the confusions that are common in many recent discussions of consumption and the environment.

The definition can be distilled from some images that commonly appear in discourse about U.S. consumption and the environment: dumps filled with disposable products, plastics, and consumer packaging waste; freeways clogged with traffic that pollutes the air but barely moves; automobiles and appliances junked when they might have been repaired; tracts of large, single-family homes with few occupants, but centrally air conditioned and with heated or cooled swimming pools; advertisements for products that no one seemed to want a few years ago but that soon everyone will "need"; air-conditioned shopping malls surrounded by acres of asphalt; and trash-lined streets and highways. In some of these images, consumers appear as acquiring and disposing of things they want but do not necessarily need; in others, they are running on a treadmill, sacrificing time with their families and friends to work increasingly long hours for money to buy things they feel they need but do not really want. The images portray excesses of resource use, waste, and material acquisitiveness and lives that are driven by, but ultimately unfulfilled by, material things. These images connote what participants in a recent U.S. study most often referred to as "materialism"—a set of values that places material abundance ahead of all else (Harwood Group, 1995).

It should not go without saying that these images embody a normative critique of consumerist culture, based on claims that it is destructive environmentally, and in some versions of the critique, destructive socially and spiritually as well (see, e.g., Institute for Philosophy and Public Policy, 1995). Many writers on consumption and the environment believe there is *too much* consumption in the United States and that people should want (or should be influenced) to consume less. Interestingly, a substantial minority of Americans say they would like to earn and consume less, especially if doing so would free more time for family life and other nonmaterialist pursuits (e.g., Schor, 1991; Harwood Group, 1995).

But leaving aside normative content, what does consumption mean in this discourse? The implicit definition might be stated this way: *Consumption consists of the purchase decisions of households and what they do with their purchases. Its environmental impacts are the transformations of materials and energy that ultimately result from these activities.* This definition embodies some assumptions about what causes the "environmental impact of consumption"—each of them heuristically useful to a point, but analytically flawed—and implies a research agenda. I first state the assumptions and then assess them and their research implications.

Assumption 1: Individuals and households are the actors most responsible for the environmental impacts of consumption. This assumption points to the importance of research on individual and household behavior, particularly consumer behavior. It draws attention away from the activities of firms and governments, except as their activities serve or induce consumers' desires.

Assumption 2: Affluence, or more specifically, affluence U.S. style, is the pattern of living through which households cause environmental impacts. This assumption implies that research should focus on the extent to which households spend (rather than save) their incomes and on their patterns of spending, particularly the extent to which spending is on materials- and energy-intensive products and services. It also suggests research on consumer decisions to forego earning (and therefore consuming) in favor of other uses of their time.

Assumption 3: The driving forces of anthropogenic environmental change, other than population growth, are economic growth and a set of forces acting on consumer "preferences." This assumption underlines the importance of research on the causes of growth in consumers' incomes. It also directs attention to such forces as individuals' values and worldviews as they concern material goods, social norms and interpersonal influences regarding material possessions, the socialization of materialist values or "consumer culture," and market-related forces affecting consumer behavior, including pricing and advertising of materials- and energy-intensive products. This assumption would direct economists toward further study of how income drives consumption. Psychologists would study factors within individuals, such as values, attitudes, knowledge, and purportedly fundamental human tendencies such as selfishness and the desire for status. Sociologists would study forces such as advertising, status competition, and the ideology of mastery over nature that came to ascendence in Western societies with the rise of capitalism.

Assumption 4: The environmental impacts of consumption are more or less the same for all kinds of consumption. This assumption, unreasonable when made explicit, in fact underlies some popular writing on consumption and the environment. Those who define consumption in ways like the above often fail to distinguish between consumption that does and does not leave the transformed materials available for reuse (e.g., lead in automotive batteries vs. lead in paint) or between consumption of things of equal price that differ in the environmental consequences of producing and using them.

This definition of consumption in terms of household behavior provides a useful heuristic as far as it goes because it points to a coherent and pertinent set of empirical questions: What causes household income to increase? What drives rates of saving? What determines the energy- and materials-intensiveness of household spending and the use of household technologies? What policies can induce households to use their incomes in more environmentally benign ways? There is room for all the social and behavioral sciences and for specialists in technology to do useful work on these questions. Moreover, answers to them would be valuable for modeling environmental change and perhaps for stopping or slowing undesirable change.

Nevertheless, the research agenda does not do justice to the issue of "environmental impacts of consumption," and the assumptions are analytically flawed. The first assumption, that most consumption is directly caused by individuals and households, is simply incorrect for the United States and other affluent countries. The vast majority of energy use, releases of water and air pollutants, and many other environmentally destructive activities in the United States results directly from organizational behavior rather than individual behavior—specifically, from the acts of corporations and governments (Gardner and Stern, 1996; Allen, Chapter 3). The most environmentally significant choices are not those that householders make, such as to purchase and then use consumer technologies, but the purchase and use choices of organizations, and organizational choices about how technologies that affect the environment are designed, produced, distributed, and marketed. To presume that consumers are entirely responsible for the environmental impacts of consumption is to overlook most of the phenomenon.

One might argue that household behavior is the ultimate driver because of consumer sovereignty, but that effect is indirect and incomplete. Most people normally have weak preferences with regard to the technology used to produce what they purchase. Also, the environmental impact of production processes is typically hidden from consumers when they make choices. It would therefore be an analytical mistake to overlook organizational decisions that directly affect the energy- and materials-intensity of the economy and do so somewhat independently of consumer choice, for example, by determining which products are available for purchase or which industrial processes are employed to manufacture them. It might also be a practical mistake, if the goal is to reduce environmental impact. Systematic campaigns to help consumers understand the environmental impacts of production processes may be effective. The potential of such an approach in this area is suggested by the growing demand in the United States for "organic" food products, which are marketed as environmentally superior.

The second assumption, that U.S.-style affluence is the source of environmental degradation, is better treated as a hypothesis to be analyzed than a conclusion. The focus on affluence suggests that researchers should classify patterns of living at different income levels—styles of affluence, frugality, poverty, and so forth—and compare their environmental effects. It is particularly important to learn how these patterns or styles are shaped by people's social, economic, and geographic contexts; how they change; and whether they can be materially influenced by acts of individual will or by policy. This assumption also suggests a sharp distinction between spending and saving that may not, in fact, be environmentally significant. Whether consumer saving is better for the environment than consumer spending is an empirical question, the answer to which depends on what kinds of investments are made by those who hold the savings. A focus on consumer spending thus distracts attention from the environmental impacts of investment, which are intimately tied to those of economic consumption.

The third assumption, about driving forces, is flawed because it focuses on only a subset of the relevant driving forces of anthropogenic environmental change. Most of the important driving forces fit into five categories: population growth, economic growth, technological change, political-economic institutions, and attitudes and beliefs (National Research Council, 1992). By omitting technology and institutions and the forces that shape them in turn, the third assumption rules out important lines of investigation. These neglected driving forces profoundly affect human transformations of materials and energy, and altering them provides ways of controlling the environmental impacts of human activity.

Thus, the household-based definition of consumption is not only inadequate for understanding but also inappropriate in policy terms: It unnecessarily narrows vision concerning the strategies available for changing consumption's environmental impacts. Such a definition focuses attention on households and on affluence, suggesting that to solve environmental problems, individuals and households must spend (and perhaps earn) less. Aside from the fact that this conclusion is unlikely to lead to acceptable policy options, it has major substantive problems.

One is that the conclusion may be overly pessimistic. There are effective policy strategies that do not directly target individuals and that, often by focusing on technology and institutions, accomplish desired goals more effectively and in more acceptable ways. For example, improving emissions control technology in automobiles, a policy directed mainly at manufacturers, did more to reduce urban air pollution than any politically practicable policy directed at households could have done. A broader definition of consumption might help identify such strategies and allow analysis of how much they can accomplish.

Another problem is that the focus on directly changing household behavior suggests a panoply of interventions that do not work well. Some of them provoke public resistance, like President Carter's call in 1979 to lower home heating levels in winter. Some interventions are too limited in scope because households are not the main cause of the targeted environmental problems. And others are likely to fail because household behavior is multiply determined and the interventions target only a single element of it, such as consumerist values or a presumed lack of information on how to cut back. Changing consumer behavior directly is a viable policy strategy, but success depends on addressing the complexities of environmentally relevant household behavior and usually requires addressing several barriers to change simultaneously (Gardner and Stern, 1996).

TOWARD A WORKING DEFINITION OF CONSUMPTION

At this stage of development of research on the environmental impacts of consumption, a working definition of consumption should not foreclose research on significant actors, major driving forces, their interrelationships, or the various possible ways to control consumption's impacts. I propose the following definition for consideration: *Consumption consists of human and human-induced transformations of materials and energy. Consumption is environmentally important to the extent that it makes materials or energy less available for future use, moves a biophysical system toward a different state or, through its effects on those systems, threatens human health, welfare, or other things people value.*[1] One might say that this is a definition of *environmental consumption*, as distinct from, for instance, economic consumption. A few points that are implicit in this definition should be stated explicitly:

(1) Consumption in this sense is not solely a social or economic activity but a human-environment transaction. Its causes (driving forces) are largely economic and social, at least in advanced societies, but its effects are biophysical. The study of consumption therefore lies at the

---

[1]When consumption makes materials and energy less available for future use, it may affect the environment in various, sometimes contradictory, ways. First there is a straightforward resource-depletion effect resulting from the fact that resources use requires energy and produces waste. Because easily accessible resources tend to be exploited first, each additional unit of the same resource tends to take more energy to extract and to produce more waste. Consumption of materials and energy can have some countervailing effects as well. For instance, a resource (e.g., natural gas) may be used in increasing amounts as a substitute for a more environmentally damaging one (e.g., oil or coal), resulting in a net environmental improvement. Also, resource depletion with its increasing economic and environmental costs may spur the development and adoption of more environmentally benign substitutes (e.g., passive solar building design), with the result that short-term environmental damage leads to long-term improvement. Because of these countervailing effects, the net long-term environmental effect of materials and energy consumption is not easily determined: careful empirical analysis is required that looks at the larger social and economic system in which resource use is embedded. I am indebted to Joel Darmstadter for emphasizing these complexities.

interface of the social and natural sciences, and seems to require their collaboration.

(2) Consumption is defined by biophysical categories such as coal and carbon dioxide, forests and croplands, rather than by social categories such as money or status. It follows that the appropriate units for measuring consumption are physical and biological, rather than economic or social (e.g., Odum, 1971; Cleveland et al., 1984; Wernick, Chapter 3).

(3) Although consumption is defined by biophysical categories, its environmental impacts are seen through human eyes. Changes of state in biophysical systems are all environmental consequences, but not all are necessarily negative from human perspectives. The connection from environmental "change" to environmental "disruption" or "harm" is mediated by human values, and individuals may disagree about whether a particular environmental effect of consumption is one to be avoided.

(4) Consumer behavior is environmentally significant consumption according to this definition only to the extent that it has environmental effects. Thus, consumer behavior may be more or less environmentally consumptive, and some of it comes close to not consuming at all. The purchase of automotive fuel is highly consumptive; by contrast, the purchase of computer software and time on the Internet are among the least consumptive of consumer activities, especially on a per-dollar basis. Similarly, producers' economic activity may be highly consumptive, or it may not (e.g., National Research Council, 1994). Waste clean-up is undertaken to reduce the environmental impact of other economic activity, and some activities of economic producers can even reduce net consumption in the environmental sense. This can happen when a firm finds ways to use its own or another firm's wastes as an input to production: the environmental damage from waste disposal and the extraction and processing of virgin materials decreases, and economic output increases.

(5) "Consumption" is not affected only by those who are consumers in the economists' sense. Producers and distributors transform materials and energy. Consumption is also affected by those whose actions indirectly shape the purchase of consumer goods and services, for example, by setting building codes or standards for the manufacture of equipment. Public officials are also responsible for large amounts of environmental consumption. Perhaps the most extensive public consumption is by military organizations, which use large amounts of fuel, metals, explosives, and the like and engage in large-scale transformation of ecosystems, especially in wartime. The actions of military and civilian public officials can have major environmental effects. For instance, by changing their purchasing practices, they can affect the environment both directly and through their influence on the producers of what they purchase.

(6) Consumer goods are not the only things that consume resources

and have environmental impacts. Public sector activities, services, and even investment are environmentally important consumption if they have major environmental consequences. Activities outside the market (e.g., religious ritual) can also transform materials and energy.

(7) All human beings and societies, not just the affluent ones, consume. The drastically different quantities and qualities of consumption around the world are a matter for empirical investigation rather than for polemic.

The broad definition of consumption has the advantage that it does not foreclose the study of human choices and activities that may hold keys to reducing the environmental impacts of human activity. The following list suggests some of the social phenomena that tend to be overlooked under the narrow, popular definition of consumption but are included in the broader one, and that may be environmentally important.

- *Changes in the structure of production and work.* The environmental impact of human choices and actions can decrease without change in households' preferences, incomes, or well-being if products are manufactured in less environmentally destructive ways (e.g., encyclopedias on line instead of on paper, fiber optic telephone lines instead of copper), and if working conditions put less stress on the environment (telecommuting may be an example). What trends move the economy in these directions? Which structural changes are environmentally beneficial? Which policies promote or hinder these changes?
- *Substitution of services for products.* Most consumer purchases are motivated by a desire for a service or function rather than for a product in itself. People buy natural gas for heating but might get some home heat from passive solar housing design; they buy automobiles to travel but might get some of this service from well-designed mass transit. People also substitute restaurant food for home cooking, a change that may or may not, on balance, be good for the environment. What drives such trends? What are their environmental implications?
- *Changes in household composition and patterns of life.* Recent sociodemographic trends such as decreasing household size, the aging of populations, and increasing labor force participation among women of child-bearing ages may have significant environmental implications. They affect demand for travel, space heating and cooling, and various other consumer services independently of any change in basic values or attitudes about the environment. Even though there may be no viable policies to change these trends, there may be ways to change their environmental implications if their effects on consumer behavior were better understood.
- *Change in nonenvironmental policies.* It is commonly asserted that

the home mortgage tax credit and the interstate highway system have indirectly increased U.S. energy demand and disrupted ecological systems. International trade policies have also been claimed to have environmental impacts. To what extent are such claims accurate? What other nonenvironmental policies shape the environmental consequences of human choices and actions, and how can such effects be estimated in advance?

These examples suggest how adopting the broad definition of consumption may have significant advantages for guiding future research. A research agenda that illuminated the above phenomena and that also addressed the more household-focused questions about the environmental implications of income growth, consumerist ideology, personal values and preferences, and the like would provide much of the knowledge needed to understand and reduce the "environmental impacts of consumption" in the United States and elsewhere.

However, the broad definition of consumption may not be entirely satisfactory because under it, "environmental impacts of consumption" seems to be a redundancy. If consumption consists of materials and energy transformations, it automatically has environmental implications, even if the effects are not necessarily undesirable (i.e., perceived as impacts). The definition makes it necessary to speak of the "environmental impact of human choices and actions" (rather than of consumption) as the object of research. The relevant field of study, then, is human choices and activities that alter the biophysical environment, especially in ways widely considered undesirable. Many of these choices and activities are those of wealthy individuals and households, as the narrower definition of consumption presumes. But the broader definition may lead researchers in productive directions that might be missed if research looks mainly at "consumers."

The broad definition may also be unsatisfying to some because it fails to point to policy goals for "reducing the negative global environmental impact of consumption of goods and services," that is, for achieving one of the central objectives of sustainable development. In particular, it does not single out affluent consumers or consumerism as a source of environmental problems but instead leaves their roles an open question. This circumspection is actually an advantage for the purpose of informing public decisions for two reasons. First, it appropriately reflects the state of knowledge. For instance, evidence suggests that the relationship between affluence and environmental degradation is not monotonic (see, e.g., Dietz and Rosa, Chapter 4). A definition that presumes that consumer behavior lies at the root of the environmental effects of human activity in the richer countries begs a question that is still open and draws attention away from potentially effective strategies for addressing those effects without restricting aggregate consumer activity.

The second rationale for a definition that does not presume the targets of policy is that choosing such targets is inherently value laden. The definition directs research attention to human choices and actions that change biophysical systems but leaves open the question of which changes in which systems are most to be avoided. The definition emphasizes that environmental changes are more or less important depending on what people value. This formulation makes an appropriate distinction between analytical questions about the effects of human activities on biophysical systems and questions about the social meaning of those effects. Although the definition does not imply a policy strategy for sustainable development, it does point a way to get the understanding needed to inform policy debates.

ACKNOWLEDGMENTS

I express gratitude to those who offered helpful comments on an earlier draft, particularly Thomas Dietz, Emily Matthews, Eugene Rosa, Vernon Ruttan, Robert Socolow, James Sweeney, and Richard Wilk.

REFERENCES

Campbell, C.
    1987   *The Romantic Ethic and the Spirit of Modern Consumerism.* London: Blackwell.
Cleveland, C.J., R. Costanza, C.A.S. Hall, and R. Kaufman
    1984   Energy and the U.S. economy: A biophysical perspective. *Science* 225:890-897.
Gardner, G.T., and P.C. Stern
    1996   *Environmental Problems and Human Behavior.* Boston: Allyn and Bacon.
Harwood Group
    1995   *Yearning for Balance, July 1995: Views of Americans on Consumption, Materialism, and the Environment.* Takoma Park, Md.: Merck Family Fund.
Institute for Philosophy and Public Policy, University of Maryland
    1995   *The Ethics of Consumption.* Special issue of *Report from the Institute for Philosophy and Public Policy* 15(4).
National Research Council
    1992   *Global Environmental Change: Understanding the Human Dimensions.* P.C. Stern, O.R. Young, and D. Druckman, eds. Committee on the Human Dimensions of Global Change. Washington, D.C.: National Academy Press.
    1994   *Assigning Economic Value to Natural Resources.* Papers presented at the Workshop on Valuing Natural Capital for Sustainable Development, July 1993. Washington, D.C.: National Academy Press.
Odum, H.W.
    1971   *Environment, Power, and Society.* New York: Wiley-Interscience.
Samuelson, P.A., and W.D. Nordhaus
    1989   *Economics,* 13th ed. New York: McGraw-Hill.

Schor, J.
   1991   *The Overworked American: The Unexpected Decline of Leisure.* New York: Basic
          Books.
Scitovsky, T.
   1992   *The Joyless Economy: The Psychology of Human Satisfaction.* New York: Oxford.
Veblen, T.
   1899   *The Theory of the Leisure Class.* New York: Macmillan.
Vitousek, P.M., P.R. Ehrlich, A.H. Ehrlich, and P.A. Matson
   1986   Human appropriation of the products of photosynthesis. *BioScience* 36:368-373.

3

Tracking the Flows of Energy and Materials

INTRODUCTION

Consumption becomes environmentally important because of the manner or extent to which it transforms materials and energy. Therefore, to understand the environmental impacts of consumption, one must understand anthropogenic changes in the flows of materials and energy. This chapter presents four brief reports, taken from presentations at the workshop, that track flows of energy and environmentally important materials or propose methods for tracking them. These reports suggest what can be learned by following materials and energy flows. Their bibliographies point to other related work.

Iddo Wernick analyzes aggregate and per-dollar materials flows within the United States, using weight and volume as indicators. Although these units are not always good proxies for environmental impacts, the analysis provides a first approximation to importance by showing which human-environment interactions are the largest; by identifying trends, it highlights the materials that are likely to be increasing or decreasing as environmental problems. For instance, many materials used in bulk, such as steel and wood, are becoming less important aspects of economic activity, and special-purpose materials used in lesser quantity, such as special alloys, plastics, and coated papers, are becoming more important (see Larson et al., 1986). The new materials have quite different environmental impacts from one another. The use of paper, despite the information revolution, has continued to increase in absolute terms and has held steady on a per-gross-national-product (GNP) basis throughout this century. This sort of analysis, combined with information on the per-unit environmental impacts of the production and consumption of particular materials, can suggest which kinds of consumption are likely to remain, or to become, environmentally important.

David Allen's report focuses on wastes, including air pollutants as well as solid wastes. Allen identifies the sources of these wastes by type of industry. He also illustrates, with an analysis of the inputs and wastes associated with producing a kilogram of polyethylene, how the environmental impacts of particular materials or energy transformations can be examined. Data like these can be combined with production data and estimates of the toxicity of each type of emission to yield comparative quantitative assessments of the environmental significance of each product of the chemical industry or some other segment of the economy. This sort of analysis can clarify the relative environmental importance of different kinds of economic activity.

David Allen's approach is similar to one that has been used in energy analysis since the 1970s. Applying the approach to materials is more difficult, however, because materials differ qualitatively to the extent that it is not always meaningful to convert them to a common unit such as joules or kilograms. In addition, unlike energy, which dissipates as waste heat, many materials need to be tracked even after they are "used," because they continue to be transported through the environment and may reappear in environmentally significant ways.

Lee Schipper's report disaggregates one class of environmentally important consumption. Schipper looks in detail at travel, a significant and growing factor in fossil energy use and associated climatic change and pollution. He disaggregates changes over time in carbon emissions from motor vehicles in wealthy countries into those attributable to levels of activity (passenger-kilometers traveled), energy intensity (fuel per passenger-kilometer), and the fuel-use characteristics of the vehicles, and then to finer levels of detail. For instance, he examines activity levels as a function of such variables as numbers of vehicles, load factors, and number and average distance of trips. He finds that while fuel consumption for travel was leveling off in the United States, largely because of decreases in fuel used per vehicle-kilometer between 1973 and 1991 (but not thereafter), this trend was not observed in other Organization for Economic Cooperation and Development (OECD) member countries: in all the countries studied, levels of activity have continued to increase since the 1970s with no sign of saturation in any country. The higher level of automobile travel in the United States is attributable to a greater number of trips of about the same average length as in other OECD countries. Schipper also examines such factors as sex and age of drivers. This sort of analysis is useful for separating technological influences on environmentally significant consumption from behavioral ones; projecting the likely environmental impacts of travel as a function of changes in incomes, age distribution, household composition, labor force participation, and other variables; and estimating the effects of policies to reduce emissions, such as fuel taxes, on different kinds of drivers.

Faye Duchin explores the possibility of developing a classification system for households, akin to the Standard Industrial Classification system, as a way to facilitate disaggregating consumption by household type and by activities within these types, and to make possible systematic study of issues of consumption and "lifestyle" such as those raised in Schipper's work. Noting that various market research firms have developed household classifications for short-term marketing purposes, Duchin suggests that a similar form of classification might be useful for detailed analysis of household consumption, including environmentally significant consumption. She notes that developing classification schemes for different countries may help illuminate the kinds of broad changes in household consumption patterns occurring in developing countries.

All four reports illustrate the potential value of tracking and disaggregating environmentally significant human activities. Such efforts advance understanding by clarifying which actions and which actors are most responsible for particular environmental changes—and which make little difference. The results of such classifications could suggest where the greatest potential lies for reducing the environmental impact of human activity. In addition, by identifying some of the immediate purposes for which people undertake activities that cause environmental damage (e.g., travel to work), this sort of research can identify sources of resistance to policy interventions and thus alert policy makers to challenges facing their efforts to reduce the environmental impacts of consumption.

REFERENCE

Larson, E.D., M.H. Ross, and R.H. Williams
  1986   Beyond the era of materials. *Scientific American* 254 (June):34-41.

CONSUMING MATERIALS: THE AMERICAN WAY[1]

Iddo K. Wernick

I focus in this paper on characterizing consumption by providing an account of all the physical materials consumed in the United States and a framework for assessing the relative scales and environmental relevance of that consumption. Assessing the materials consumption of a nation requires viewing (a) the total volume of materials consumed, (b) the composition of that total, (c) how these change with time, (d) forces driving those changes, (e) foreign trade in raw materials, and (f) the prospects for large-scale materials recovery. Together, these allow us to view materials consumption comprehensively and place particular instances and anecdotes in proper perspective.

Since the oil price shocks of the 1970s many have studied energy consumption at the national level, examining consumption trends over time, the mix of fuels used, and alternatives for the future (United Nations, 1978; World Energy Conference, 1974). Such studies provide the analytic tools that have documented the slowing growth of primary energy consumption and its decoupling from U.S. economic development (Nakićenović, 1996). Although the analogy is imperfect, materials would similarly benefit from this approach but have not yet enjoyed the same scrutiny for several reasons. The same fear of imminent shortages that focused attention on energy never gathered momentum with respect to materials, as the proven reserve base for most material resources has actually grown in the last decades (U.S. Congress, 1952; Goeller and Weinberg, 1976; World Resources Institute, 1994). Although exhausting materials resources may, in fact, not be a priority concern—with the notable exception of high-grade energy fuels—the environmental degradation resulting from extracting, processing, consuming, and disposing materials is.

The heterogeneity of materials consumed in modern society presents a further barrier to comprehensive analysis. Materials possess numerous and diverse properties that make them attractive to consumers and determine their environmental impacts, thus weakening generalizations. While the energy from firewood, coal, or gas is readily reduced to common units such as joules or British thermal units, the utility of the gravel, ore, and fertilizer materials we consume cannot be.

Although less than an ideal measuring stick, mass will serve here as

---

[1]A longer version of this paper appeared in *Technological Forecasting and Social Change*, 1996, 53:111-122. Printed with permission of the journal.

the common currency for describing materials. Using mass alone obscures environmentally important features of materials use, such as the growing volume of plastics in U.S. waste streams, the high toxicity of relatively trivial masses of industrial effluent, and the acreage disturbed in extracting both renewable and nonrenewable resources. Nonetheless, kilograms and tons provide objective measures for grasping the sheer quantities of bulk materials mobilized to serve society and the relative sizes of different materials classes. Moreover, most of the available data on materials are either given directly in mass or can be converted to it.

CURRENT NATIONAL MATERIALS CONSUMPTION AND TEMPORAL DYNAMICS

In 1990, the average American consumed over 50 kg of material per day, excluding water (Wernick and Ausubel, 1995). Consumer goods compose a small fraction of this total; the materials required for their production and distribution, as well as the auxiliary materials used in their manufacture, contribute a far greater amount. To gain some perspective on the ratio of direct to indirect consumption, the mass of municipal waste that Americans directly dispose of each day accounts for less than 5 percent of the daily quantity (Franklin Associates, Ltd., 1992). Figure 3-1 shows the total as a sum of the six major classes of materials. Almost 90 percent of total inputs go to providing energy, structures, and food. Inputs of water, if included, would raise the total many fold. Mining wastes (particularly for coal) are huge and represent another consequence of consumption mostly hidden from the public eye. The daily 50-kg quantity may be common to highly industrialized societies. In 1990, Japanese consumption also summed to a little over 50 kg per capita per day (Gotoh, 1994).

The mix of materials consumed changes over time. For example, per capita U.S. lumber consumption has declined markedly in this century. At the turn of the century wood provided building materials for homes and factories, ties and rolling stock for railroads, utility poles for telephone and power lines, and fuel. Today a large fraction of harvested wood (approximately 40 percent including residues) goes to paper mills (Ince, 1994). Although drastic reductions in consumption are more the exception than the rule, wood is not unique in that both the level of consumption and how it is used in the economy have changed.

A more aggregated account of consumption reveals wholesale changes in the amount of physical structure materials Americans consume. Figure 3-2 shows that in total tonnage per capita, reported consumption appears to rise over long cycles of economic growth and then to fluctuate during times of economic upheaval.

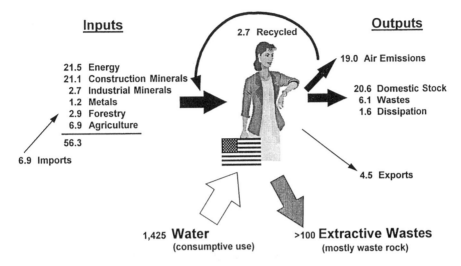

FIGURE 3-1 Daily per capita materials flow by mass (all values in kg): United States about 1990. Materials are here classed as energy fuels (i.e., coal, oil, gas), construction minerals, industrial minerals, metals, forestry products, and agricultural products. Data from Wernick and Ausubel (1995). Reprinted with permission.

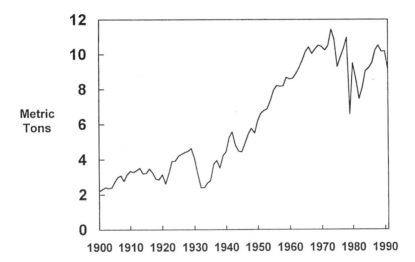

FIGURE 3-2 Annual per capita consumption of physical structure materials: United States 1900-1991. Physical structure materials are here defined as construction minerals, industrial minerals, forestry products. Data from Rogich et al. (1993); U.S. Bureau of the Census (1975). Reprinted with permission.

Are industrialized societies constrained to follow this path indefinitely? Do improvements in the standard of living necessarily translate to greater material consumption? Intensity of use (IOU) measures address this question directly. IOU measures show the evolution of individual materials used in the national economy by indexing primary, as well as finished, materials to gross domestic product (Malenbaum, 1978). Beginning with studies done in the late 1970s, researchers noted several common patterns in the course of consumption of a material in the economy (Williams et al., 1987). Initially, the consumption of a particular material exceeds general economic growth. Growing markets and newly discovered uses for the material stimulate further growth. This rapid growth eventually saturates, and consumption of that material then tracks or lags the rest of the economy.

Figure 3-3 illustrates this phenomenon at different stages for a variety of materials in the United States. One clear conclusion from the figure is that more dollars in the economy do not always mean more tons. Heavy materials such as steel, copper, lead, and lumber, all materials used for infrastructure, became less critical to economic growth over the course of this century. Paper seems to track economic activity in lockstep, conserv-

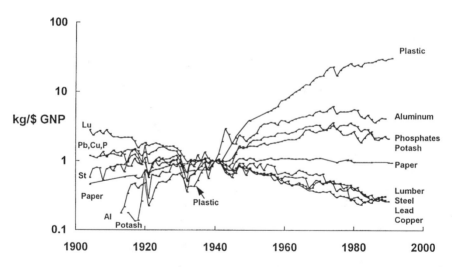

FIGURE 3-3 Materials intensity of use: United States 1900-1990. Annual consumption data are divided by GNP in constant 1987 dollars and normalized to unity in the year 1940. Data for plastics are production data. NOTE: St: Steel; Lu: Lumber. Data from U.S. Bureau of the Census (1993-1994 and 1975); *Modern Plastics* 136(5):71-72(1959); plastics data from personal communication with Joel Broyhill, Statistics Department, Society of the Plastics Industry, Washington, D.C., August 20, 1993. Reprinted with permission.

ing its role through the national shift from manufacturing to information and services. The rapid growth of materials used as fertilizers shows the "green" revolution that has raised agricultural yields. Finally, low-density materials, such as aluminum, have outpaced economic activity in the second half of the twentieth century. This is spectacularly true with respect to plastics, a class of materials that, in addition to being lightweight, possess a host of properties that make them the material of choice for the manufacturer and the consumer alike.

The types of material flows can be separated into the categories of elephants and fleas. Some of the bulk materials we have seen may be called the elephants. These high-volume material flows may cause little environmental impact per unit mass but can have profound long-range environmental consequences. Pumping oil, quarrying stone, and harvesting feed each contributes to chronic global environmental problems, affecting atmospheric composition and land use. The fleas, materials generated in small quantities often as by-products of large-scale commercial production, can have more acute harmful effects. Consider that total annual U.S. dioxin releases are under 500 kg (Thomas and Spiro, 1994). Despite the small quantity released, environmental concerns about the effects of dioxin continue to demand the attention of both government and industry. According to the U.S. Environmental Protection Agency's inclusive definition of Toxic Release Inventory (TRI) production-related wastes, toxic chemicals totaled about 17 million metric tons in 1992, 0.3 percent of all materials consumption (INFORM, 1995). Concerns over this relatively small mass fraction dominate much of the current public environmental debate.

Foreign trade in raw materials accounts for about 10 percent of U.S. materials flows. Table 3-1 shows that a few bulk commodities dominate trade. On a mass basis, agricultural products, coal, and chemicals dominate U.S. exports, whereas oil, oil products, and metals and ores dominate imports. The plentiful carbon that enters America, of course, exits as CO_2 emissions. Agricultural trade surpluses require domestic land, chemicals, and minerals but feed many elsewhere. For many minerals the United States shall continue to rely on foreign sources.

FORCES AFFECTING MATERIALS CONSUMPTION

The simple arithmetic of a U.S. population of 400 million or more in 2100 will draw more materials into the economy (United Nations, 1992). Efficiency improvements might be able to maintain a constant total for the collective whole, in theory. However, in the United States more people means more individual consumers acting on their own. The average number of residents per American occupied housing unit halved since

TABLE 3-1 Major Materials Flows in U.S. Foreign Trade

Category	Exports (million metric tons)	Imports (million metric tons)	Net Flow per Capita (kg)
Agricultural products	135.5	14.9	(482.6)
Coal	96.0	2.4	(374.5)
Minerals	47.8	54.2	25.6
Metals and ores	27.0	76.4	197.8
Chemical and allied products	41.3	14.4	(107.6)
Petroleum products	34.1	96.9	251.3
Timber products	16.4	18.4	8.0
Paper and board	6.2	11.9	22.8
Oil (crude)	5.6	307.4	1207.6
Natural gas	1.7	31.0	117.2
Automobiles[a]	1.2	5.9	18.8
TOTAL	412.7	633.8	884.4
Air transport	1.5	1.7	0.8
Waterborne transport	406.9	524.9	472.0
Trucks	151,000 (units)	766,000 (units)	N.A.
Other industrial and consumer products	?	?	

NOTE: N.A. indicates not applicable. Numbers in parentheses indicate net exports.

[a]Based on an estimated average vehicle mass of 1.5 metric tons.

SOURCE: U.S. Bureau of the Census (1975).

the beginning of the century (U.S. Bureau of the Census, 1975, 1993, 1994). Besides the materials needed for additional structures, appliances and furniture enter these dwellings irrespective of the number of inhabitants. Thus the relationship of number of persons to materials consumed is not simply proportional, reflecting settlement patterns as well. This same relation holds true for energy consumption: the same number of people living in a larger number of residences consume more (Schipper, 1996).

While American behavior drives expansion, historical development and technical innovations offer hope for contraction. The United States is a postindustrial country. The service sector continues to claim more of national economic activity, and the physical infrastructure of the country is largely in place. For instance, during the period 1970-1992, the surfaced road network in the United States expanded at only a third of the rate for the century (U.S. Bureau of the Census, 1975, 1993, 1994). Because of the

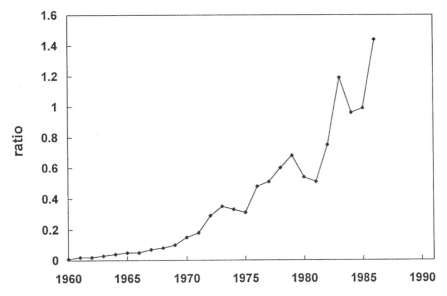

FIGURE 3-4 Volume ratio of pipe manufactured from plastic over all other materials. Data from Hurdelbrink (1989). Repinted with permission.

massive quantities consumed constructing roads and highways, slowing the rate is consequential to national consumption of materials like steel, asphalt, sand, and rock.

Substituting lighter for heavier materials also puts downward pressure on national materials use. Replacing heavy copper cable with light fiber optics not only reduces the amount of mass consumed but also reduces the need for mining copper ore. Lightweight plastics now provide the primary material for pipes, formerly made of steel and lead (Figure 3-4). The quantity of carbon steel in American automobiles fell drastically during the 1970s, while high strength steel alloys, plastics, composites, and aluminum continue to make up more of our cars (Figure 3-5).

For some products the same utility can be supplied with less mass of product. Metallurgical advances allow for steel beams with smaller cross-sectional areas to support loads. Sweetening foods with high-fructose corn syrup uses only a fifth the mass of sugar to produce the same result to our palate. The ubiquitous aluminum beverage can is today 25 percent lighter than in 1973 (personal communication, Jenny Day, Director of Recycling, Can Manufacturers Institute). In addition to smaller mass, the aluminum beverage can provide a model of a highly successful recycling system with a recycling rate exceeding 70 percent.

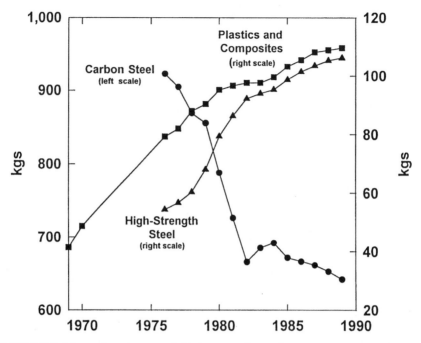

FIGURE 3-5 Mass of carbon steel, high-strength steel, composite materials, and plastics in the average U.S. automobile: 1969-1989. Data from *Ward's Automotive Yearbook* (1970–1989). Reprinted with permission.

The combination of forces to reduce materials use in the industrialized countries drives a process that researchers have dubbed "dematerialization," or aggregate reductions in the amount of material needed to serve economic functions (Wernick et al., 1996). Substitution of materials that require less mass to deliver a unit of a given service, a phenomenon formally named "transmaterialization," represents a central component of the proposed shift to lowered consumption.

Developing nations can benefit from the knowledge-based shift to lower materials requirements. The dematerialization hypothesis maintains that as nations launch into development later, their initial growth rates may be sharper, but consumption saturates at lower levels, as they can avoid the materials-intensive process of trial and error experienced by the earlier starters (Grubler, 1990). The potential for reducing materials use through recycling, or "materials recovery," can also be studied in terms of mass transformations (Rogich, 1993; Wernick, 1994; Wernick et al., 1996; Allen, this chapter).

CONCLUSIONS

Sustaining the U.S. economy requires consuming large amounts of materials. The mix of materials changes with time, and these changes matter from the perspective of environmental quality. The question of whether Americans will consume more or less materials in the future depends on demographic, economic, and technical variables difficult, if not impossible, to predict. One central question is whether increases in materials efficiency can keep pace with, or even triumph over, the forces driving increased consumption.

Toxics and other harmful materials constitute a small part of total consumption but are currently linked to the large-scale production of goods. They pose threats to human health and environmental quality far exceeding their mass fraction of materials consumption. To what extent these nasty residuals, often unintended by-products of production, can be eliminated presents a further question.

The demand for better performance, and hence greater sophistication in materials and goods, has lightened many products and is key to future trends in materials consumption and efforts in materials recovery. Research and development efforts must combine environmental objectives with consumption trends to reduce primary materials requirements, design products for recovery, and find uses for so-called wastes.

While technology may offer some solutions and help reduce the environmental impact of our consumption, changing human behavior will surely prove more difficult. Technological and economic solutions must recognize the deep behavioral forces driving human consumption to effect positive change.

ACKNOWLEDGMENTS

I thank Jesse Ausubel and Perrin Meyer at The Rockefeller University and Paul Waggoner at the Connecticut Agricultural Experiment Station for their help in preparing this manuscript.

REFERENCES

Franklin Associates, Ltd.
  1992 *Characterization of Municipal Solid Waste in the United States: 1992 Update, Final Report.* EPA/530/R-92/019. Prairie Village, Kan.: Franklin Associates, Ltd.
Goeller, H.E., and A.M. Weinberg
  1976 The age of sustainability. *Science* 191:683-689.

Gotoh, S.
  1994   The Potential and Limits of Using Life-Cycle Approach for Improved Environ-
          mental Decisions. Paper presented at the International Conference on Industrial
          Ecology sponsored by the National Academy of Engineering. May 9-13, Irvine,
          Calif.
Grubler, A.
  1990   *The Rise and Fall of Infrastructures: Dynamics of Evolution and Technological Change
          in Transport.* Heidelberg, West Germany: Physica-Verlag.
Hurdelbrink, R.
  1989   An analysis of materials selection criteria for synthetic polymer systems. In *Pro-
          ceedings of the Industry-University Advanced Materials Conference II.* Denver, Colo.:
          Advanced Materials Institute, Colorado School of Mines.
Ince, P.J.
  1994   Recycling of Wood and Paper Products in the United States. Paper presented at
          United Nations Economic Commission for Europe Timber Committee Team of
          Specialists on New Products, Recycling, Markets, and Applications for Forest
          Products. June 1994.
INFORM
  1995   *Toxics Watch 1995.* New York: Inform, Inc.
Malenbaum, W.
  1978   *World Demand for Raw Materials in 1985 and 2000.* New York: McGraw-Hill.
Nakićenović, N.
  1996   Freeing energy from carbon. *Daedalus* 125(3):95-112.
Rogich, D.G.
  1993   Materials Use, Economic Growth, and the Environment. Paper presented at the
          International Recycling Congress and REC'93 Trade Fair. Washington, D.C.: U.S.
          Bureau of Mines.
Schipper, L.
  1996   Life-styles and the environment: The case of energy. *Daedalus* 125(3):113-138.
Thomas, V.M., and T.J. Spiro
  1994   *An Estimation of Dioxin Emissions in the United States.* PU/CEES Report No. 285.
          Princeton, N.J.: Center for Energy and Environmental Sciences, Princeton Univer-
          sity.
United Nations
  1978   *World Energy Supplies 1972-1976.* New York: United Nations.
  1992   *Long-Range World Population Projections: Two Centuries of Population Growth 1950-
          2150.* New York: United Nations.
U.S. Bureau of the Census
  1975   *Historical Statistics of the United States, Colonial Times to 1970.* Washington, DC:
          U.S. Department of Commerce.
  1993   *Statistical Abstract of the United States,* 113 ed. Washington, D.C: U.S. Department
          of Commerce.
  1994   *Statistical Abstract of the United States,* 114 ed. Washington, D.C: U.S. Department
          of Commerce.
U.S. Congress
  1952   *Resources for Freedom: Report of the President's Materials Policy Commission* (Paley
          Report). U.S. House of Representatives Document No. 527. Washington, D.C.:
          U.S. Government Printing Office.
Ward's Communications
  1970-   *Ward's Automotive Yearbook.* Detroit: Ward's Communications.
  1989

Wernick, I.K.
  1994   Dematerialization and secondary materials recovery: A long run perspective. *Journal of the Minerals, Metals, and Materials Society* 46(4):39-42.
Wernick, I.K., and J.H. Ausubel
  1995   National materials flows and the environment. *Annual Review of Energy and Environment* 20:462-492.
Wernick, I.K., R. Herman, S. Govind, and J.H. Ausubel
  1996   Materialization and dematerialization: Measures and trends. *Daedalus* 125(3):171-198.
Williams, R.H., E.D. Larson, and M.H. Ross
  1987   Materials, affluence and industrial energy use. *Annual Review of Energy and Environment* 12:99-149.
World Energy Conference
  1974   *World Energy Conference Survey of Energy Resources.* New York: World Energy Conference.
World Resources Institute
  1994   *World Resources: A Guide to the Global Environment 1994-1995.* New York: Oxford University Press.

WASTES AND EMISSIONS IN THE UNITED STATES

David T. Allen

More than 12 billion tons of industrial waste are generated annually in the United States. This is equivalent to more than 40 tons of waste for every man, woman, and child in the country [U.S. Environmental Protection Agency (E.P.A.), 1988a,b; Allen and Jain, 1992]. The sheer magnitude of the waste generation is cause for concern and drives us to identify the characteristics of the wastes, the manner in which the wastes are being managed, and the potential for reducing wastes. This summary provides a brief overview of the information available on waste generation and management. The sources and nature of industrial hazardous wastes, nonhazardous wastes, municipal solid wastes, and emissions of criteria and hazardous air pollutants will all be reviewed.

INDUSTRIAL WASTES

Industrial wastes can be in solid, liquid, or gaseous states. Most industrial wastes in solid or liquid form fall under the provisions of the Resource Conservation and Recovery Act (RCRA). The total mass of wastes generated each year that fall under the provisions of RCRA has been estimated to be in excess of 12 billion tons. Table 3-2 lists estimated rates of nonhazardous waste generation in the United States (U.S. E.P.A., 1988a). The total of more than 11 billion tons can be contrasted with an estimate of 0.75 billion tons of hazardous waste generation (U.S. E.P.A., 1991; Baker et al., 1992). Thus, hazardous wastes represent less than 10 percent of the total industrial waste mass, yet almost all of our data on waste generation and treatment focus on this small segment of the waste system. Figure 3-6 shows the industrial sectors responsible for the generation of hazardous wastes. Chemical manufacturing clearly dominates. In contrast, the paper, mining, and electric power-generation industries dominate estimates of nonhazardous waste generation.

Although they account for more than 12 billion tons of industrial waste generation, wastes regulated under RCRA do not represent the entire waste burden. Emissions to the atmosphere do not, in general, fall under the provisions of RCRA. Data on emissions into the atmosphere can be broadly classified into emissions of hazardous air pollutants and criteria air pollutants. The richest source of data on atmospheric emissions of hazardous air pollutants from industrial facilities is the Toxics Release Inventory (TRI). The TRI reports the releases and transfers of more than 300 chemicals (soon to be 600 chemicals) from manufacturing facilities. In 1991, more than 7 billion pounds of releases were reported

TABLE 3-2 Sources of Nonhazardous Waste Regulated under RCRA

Waste Category	Estimated Annual Generation Rate (million tons)
Industrial nonhazardous waste[a,b]	7,600
Oil and gas waste[c,e]	
Drilling waste[d]	129–871
Produced waters[f]	1,966–2,738
Mining waste[c,g]	> 1,400
Municipal waste[b]	158
Household hazardous waste	0.002–0.56
Municipal waste combustion ash[h]	3.2–8.1
Utility waste[c,i]	
Ash	69
Flue gas desulfurization waste	16
Construction and demolition waste[j]	31.5
Municipal sludge[b]	
Wastewater treatment	6.9
Water treatment	3.5
Very-small-quantity[k] generator	
Hazardous waste (<100 kg/mo)[b,e]	0.2
Waste tires[g]	240 million tires
Infectious waste[c,l]	2.1
Agriculture waste	Unknown
Approximate Total	> 11,387

[a]Not including industrial waste that is recycled or disposed of off site.
[b]These estimates are derived from 1986 data.
[c]See Science Applications International Corporation, 1985.
[d]Converted to tons from barrels: 42 gal = 1 barrel, ≈ 17 lb/gal.
[e]These estimates are derived from 1985 data.
[f]Converted to tons from barrels: 42 gal = 1 barrel, ≈ 8 lb/gal.
[g]These estimates are derived from 1983 data.
[h]This estimate is derived from 1988 data.
[i]These estimates are derived from 1984 data.
[j]This estimate is derived from 1970 data.
[k]Small quantity generators (100-1,000 kg/mo waste) have been regulated under RCRA, Subtitle C, since October 1986. Before then, approximately 830,000 tons of small-quantity generator hazardous wastes were disposed of in Subtitle D facilities every year.
[l]Includes only infectious hospital waste.

SOURCE: U.S. Environmental Protection Agency (1988a).

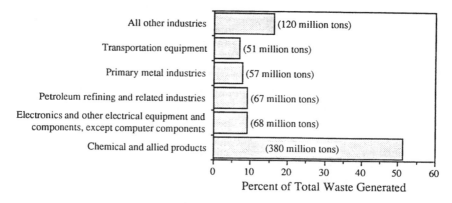

FIGURE 3-6 Industrial sources of hazardous waste. Data from Baker and Warren (1992).

through the TRI. The major industry categories contributing to this total are listed in Table 3-3. While TRI releases are more than three orders of magnitude less than the mass of wastes reported under RCRA, RCRA wastes are sent to treatment technologies that reduce waste volume and toxicity, whereas TRI releases are generally emitted directly to the environment. Thus, direct releases to the environment, reported through the TRI, may be as significant in affecting environmental quality as RCRA

TABLE 3-3 Sources of Releases and Transfers of Toxics

Industry	Standard Industrial Classifications Code	Total Releases and Transfers	
		(millions of pounds)	(% of total)
Classification			
Food and kindred products	20	85	1.2
Tobacco products	21	3.6	0.049
Textile mill products	22	37	0.51
Apparel, etc.[a]	23	1.9	0.026
Lumber and wood products, nec[b]	24	41	0.56
Furniture and fixtures	25	67	0.93
Paper and allied products	26	310	4.2
Printing, publishing, and allied industries	27	57	0.78
Chemicals and allied products	28	2,800	38
Petroleum refining and related industries	29	720	9.9
Rubber and miscellaneous plastic products	30	200	2.7
Leather and leather products	31	19	0.27

TABLE 3-3 Continued

Industry	Standard Industrial Classifications Code	Total Releases and Transfers	
		(millions of pounds)	(% of total)
Classification			
Stone, clay, glass, and concrete products	32	55	0.75
Primary metal industries	33	1,200	17
Fabricated metal products, nec[b]	34	340	4.7
Industrial and commercial machinery and computer equipment	35	91	1.3
Electronic and other electrical equipment and components, nec[b]	36	350	4.8
Transportation equipment	37	310	4.3
Measuring instruments, optical goods, watches[c]	38	71	0.98
Miscellaneous manufacturing industries	39	34	0.46
Multiple manufacturing classifications			
Stone, etc. products/primary metal industries	32 and 33	37	0.51
Primary metals/fabricated metals	33 and 34	36	0.50
Chemical products/petroleum refining	28 and 29	35	0.49
Primary metals/electronic equipment	33 and 36	26	0.35
Chemical products/primary metals	28 and 33	25	0.35
Chemical products/rubber, plastic products	28 and 30	25	0.34
Food products/chemical products	20 and 28	20	0.28
Paper products/electronic equipment	26 and 36	17	0.24
Electronic equipment/transportation equipment	36 and 37	14	0.19
Fabricated metals/transportation equipment	34 and 37	11	0.15
Fabricated metals/electronic equipment	34 and 36	11	0.15
Primary metals/machinery	33 and 35	11	0.15
Other manufacturing industry combinations	20 to 39	170	2.3
Not classified as manufacturing	No data, < 20, > 39	38	0.53
TOTAL		7,300	100

[a]Full description is as follows: Apparel and other finished products made from fabrics and other similar materials.

[b]nec: not elsewhere classified.

[c]Full description is as follows: Measuring, analyzing, and controlling instruments; photographic, medical, and optical goods; watches and clocks.

SOURCE: Toxic Release Inventory (1993).

wastes, even though the mass of RCRA wastes dwarfs the mass reported through the TRI. Criteria pollutants are the other major category of air pollutants. The criteria pollutants include particulate matter less than 10 microns in diameter, sulfur dioxide, nitrogen oxides, carbon monoxide, ozone, and lead. Volatile organic hydrocarbons are not included in the list of criteria pollutants, but because of their role in forming ozone in the lower atmosphere, they are measured and controlled. The criteria air pollutants emitted by major industrial sectors are given in Figure 3-7. The emissions are on the order of millions of tons per year, comparable to the emissions reported through the TRI, but orders of magnitude lower than the wastes governed by RCRA.

MUNICIPAL SOLID WASTES AND OTHER WASTE STREAMS

Approximately 200 million tons of municipal solid waste are generated annually in the United States. The main constituents are paper, yard wastes, food wastes, plastics, and metal. While much attention has been focused on these materials, the quantities of these waste streams are just a few percent of the overall waste material flows in the United States. Industrial wastes generated in the production of commodity materials and the generation of power and fuel are far more extensive than post-consumer materials. Agricultural wastes are also extensive but are not well characterized.

IMPLICATIONS OF WASTE STREAM FLOWS

Per capita waste generation in the United States is approximately 40 tons per year. This amount is roughly equivalent to each person generating their body weight in wastes each day when water is excluded from the 40 tons (Wernick and Ausubel, 1995). In another paper in this volume, Wernick reports that per capita material consumption in the United States is also approximately a body weight per day. Making detailed comparisons between the material use and waste generation is difficult because wastes are often poorly characterized, making the determination of flows of specific materials uncertain. Further, because the measured flows of wastes often contain significant quantities of diluting species such as water, even performing total mass balances on materials used and wastes generated is difficult. Because of these difficulties, which are described in more detail elsewhere (Allen and Jain, 1992), this paper only briefly summarizes the waste flow and comparison data and does not attempt to compare them in detail to material use data, although such comparisons should be the topic of future research.

The available data on wastes clearly indicate that most materials used

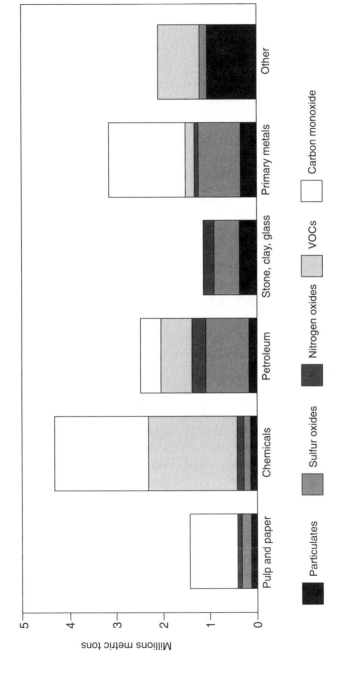

FIGURE 3-7 Industrial sources of criteria air pollutants (United States, 1994). NOTE: VOCs = volatile organic compounds. Data from U.S. Department of Energy (1994).

in the United States are from virgin resources, and only a relatively small fraction are recovered or reprocessed materials. Yet these 12 billion tons per year of wastes should not be ignored as a potential resource. Studies in what has come to be called industrial ecology or industrial metabolism are probing the material efficiencies of large industrial systems, searching for ways to improve material and energy efficiencies. Although it may be argued that low rates of material reuse are due to the inherently low value of waste streams, data on waste compositions tell a different story. For example, Allen and Behmenesh (1994) found that billions of dollars in metals are disposed of each year in hazardous waste streams, and a large fraction of this material is present at concentrations higher than those found in ores that are currently mined. The reasons for these lost opportunities are complex (Allen, 1993), but it is clear that much more material-efficient industrial systems are feasible.

ENVIRONMENTAL IMPACTS OF CONSUMPTION

There are several lessons to be learned from waste and emission inventories.

• Waste flows are substantial and are dominated by the by-products of the manufacture of commodity materials and energy; post-consumer waste flows are relatively minor in comparison.
• Many of the waste streams contain valuable resources at concentrations that should allow for economical recovery, yet a series of regulatory, financial, structural, and technical barriers make the development of more efficient structures difficult.
• Information on waste stream flows exists in many disparate locations. For some waste streams very little information is available. This lack of information is a major barrier to the formation of markets that could find uses for materials currently wasted.

To be most useful in estimating the environmental impacts of consumption, the waste and emission data should be aggregated along product lines. The emerging discipline of Life Cycle Assessment attempts to accomplish this task. Consider Table 3-4, which is a compilation of the wastes, emissions, and raw materials used in generating a kilogram of polyethylene. These data reveal that emissions and wastes are generated throughout a product's life cycle, from extraction of raw materials, through manufacturing, and in final product disposal. Further, the wastes and emissions fall into a variety of categories and will have a variety of impacts. In the face of this complexity, developing simple measures of the environmental impacts of consumption will be challenging. Despite

TABLE 3-4 Raw Material Consumption, Emissions, and Energy Use Associated with the Manufacture of 1 kg of Polyethylene

	Category	Unit Average
Fuels, MJ	Coal	3.28
	Oil	3.58
	Gas	12.38
	Hydro	0.54
	Nuclear	1.67
	Other	0.21
	TOTAL	21.66
Feedstock, MJ	Coal	< 0.01
	Oil	33.87
	Gas	33.02
	Wood	< 0.01
	TOTAL	66.89
TOTAL FUEL PLUS FEEDSTOCK		88.55
Raw materials, mg	Iron ore	200
	Limestone	150
	Water	24,000,000
	Bauxite	300
	Sodium chloride	8,000
	Clay	20
	Ferro-manganese	< 1
Air emissions, mg	Dust	3,000
	Carbon monoxide	900
	Carbon dioxide	1,250,000
	Sulfur oxides	9,000
	Nitrogen oxides	12,000
	Hydrogen chloride	70
	Hydrogen fluoride	5
	Hydrocarbons	21,000
	Other organics	1
	Metals	5
Water emissions, mg	COD	1500
	BOD	200
	Acid as H^+	60
	Nitrates	5
	Metals	250
	Ammonium ions	5
	Chloride ions	130
	Dissolved organics	20
	Suspended solids	500
	Oil	200
	Hydrocarbons	100
	Dissolved solids	300
	Phosphate	5
	Other nitrogen	10

continued on next page

TABLE 3-4 Continued

	Category	Unit Average
Solid waste, mg	Industrial waste	3,500
	Mineral waste	26,000
	Slags and ash	9,000
	Toxic chemicals	100
	Nontoxic chemicals	800

NOTE: The feedstock energy is the energy value of the fuels used as the raw materials for making polyethylene; the other fuels listed are consumed in the manufacturing and raw material extraction processes. COD: chemical oxygen demand; BOD: biological oxygen demand.

SOURCE: Boustead (1993).

the challenges, systems are emerging to perform product environmental impact assessments and these are described in a growing literature on Life Cycle Assessment (Society of Environmental Toxicology and Chemistry, 1991; 1993a,b; 1994).

REFERENCES

Allen, D.T.
   1993   Using wastes as raw materials: Opportunities to create an industrial ecology. *Hazardous Waste and Hazardous Materials* 10:273-277.
Allen, D.T., and N. Behmanesh
   1994   Wastes as raw materials. Pp. 69-89 in B.R. Allenby and D.J. Richards, eds., *The Greening of Industrial Ecosystems*. National Academy of Engineering. Washington, D.C.: National Academy Press.
Allen, D.T., and R. Jain, eds.
   1992   *Hazardous Waste and Hazardous Materials* 9(1):1-111.
Baker, R.D., and J.L. Warren
   1992   Generation of hazardous waste in the United States. *Hazardous Waste and Hazardous Materials* 9:19-35.
Baker, R.D., J.L. Warren, N. Behmanesh, and D.T. Allen
   1992   Management of hazardous waste in the United States. *Hazardous Waste and Hazardous Materials* 9:37-59.
Boustead, I.
   1993   *Ecoprofiles of the European Plastics Industry*. Reports 1-4 (May). Brussels, Belgium: European Centre for Plastics in the Environment (PWMI).
Science Applications International Corporation
   1985   *Summary of Data on Industrial Non-hazardous Waste Disposal Practices*. Contract 68-01-7050. Washington, D.C.: U.S. Environmental Protection Agency.
Society of Environmental Toxicology and Chemistry (SETAC)
   1991   *A Technical Framework for Life Cycle Assessment*. Report from the Smuggler's Notch, Vt., workshop held August 1990. Pensacola, Fla.: SETAC.

1993 *Conceptual Framework for Impact Assessment*. Report from the Sandestin, Florida, workshop held February 1992. Pensacola, Fla.: SETAC.

1993 *Guidelines for Life Cycle Assessment*. Report from the workshop held in Sesimbra, Portugal in March 1993. Pensacola, Fla.: SETAC.

1994 *Life Cycle Assessment Data Quality: A Conceptual Framework*. Report from the Wintergreen, Va., workshop held October 1992. Pensacola, Fla.: SETAC.

Toxic Release Inventory

1993 *Toxic Chemical Release Inventory for 1991* (database). Bethesda, Md.: National Library of Medicine.

U.S. Department of Energy

1994 Waste Generation in Industry, Draft, Office of Energy Efficiency, Washington, D.C.

U.S. Environmental Protection Agency

1988a *Report to Congress: Solid Waste Disposal in the United States*, Vol. 1, EPA/530-SW-88-011. Washington, D.C.: U.S. Environmental Protection Agency.

1988b *Report to Congress: Solid Waste Disposal in the United States*, Vol. 2, EPA/530-SW-88-011B. Washington, D.C.: U.S. Environmental Protection Agency.

1991 *National Survey of Hazardous Waste Generators, and Treatment, Storage, Disposal and Recycling Facilities in 1986: Hazardous Waste Management in TSDR Units*. EPA/530-SW-91-060. Washington, D.C.: U.S. Environmental Protection Agency.

Wernick, I.K., and J.H. Ausubel

1995 National material flows and the environment. *Annual Review of Energy and the Environment* 20:463-492.

CARBON EMISSIONS FROM TRAVEL
IN THE OECD COUNTRIES

Lee J. Schipper

This paper examines some of the forces driving increased emissions of greenhouse gases in developed countries, focusing on carbon dioxide and travel (a longer paper addresses other sources of emissions, Schipper, forthcoming). Figure 3-8 shows carbon emissions per capita in OECD countries in 1973 and 1991 from energy-using activities, allocating the emissions from production of electricity and district heating to the end uses of those energy forms in proportion to final use (Schipper, Haas, and Sheinbaum, 1996; Scholl et al., 1996; Schipper, Scholl, and Price, in press; Schipper, Ting et al., 1996; Torvanger, 1991). The figure illustrates well that it is possible to connect emissions to the activities where they arise. For many countries, per capita emissions from these activities actually fell between the years portrayed; normalizing by gross domestic product in each year shown would show a dramatic decline in every country. Indeed, absolute emissions for most countries shown were close to or lower than their 1973 levels in 1991, but absolute emissions have begun to rise since then.

This analysis uses Laspeyres indices to measure how components of energy use changed (Howarth et al., 1993). The analysis decomposes or factors total energy use or emissions into a sum, over end-use sectors (such as modes of travel), of the products of subsectoral activity and energy or emissions intensity (i.e., energy or emissions per unit of activity): $E = \Sigma A_i \times E_i$ where E is energy use or emissions, A_i is activity in passenger-kilometers in mode i (including the driver for automobiles and light trucks), E_i is the energy or emisions per unit of activity in mode i, and A_i is calculated by multiplying A (total passenger-kilometers traveled) by S_i (the share of activity in mode i). Letting one component follow its historic course while holding the others constant at their base year values shows how that component influenced energy use over time.

This approach (see Schipper, forthcoming) shows that the growth in energy demand is shifting from producers (manufacturing and freight) to consumers (household comfort and mobility), as well as to services; these services act as both producers (i.e., insurance, banking, health) and consumers (personal services like shopping and leisure activities). This shift "from production to pleasure" is barely discernible for the United States (Figure 3-9) because energy savings there were so great for household purposes and private cars, but it is very noticeable in Germany (Figure 3-10) and other OECD countries. The shift depends on the rising impor-

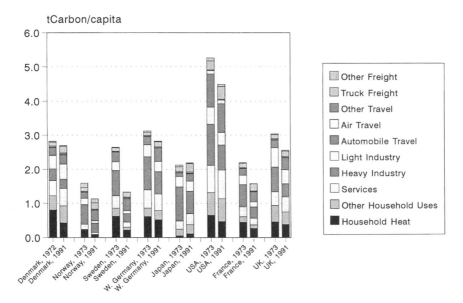

FIGURE 3-8 Carbon emissions per capita in 8 OECD countries, 1972-1973 and 1991 by end use. NOTE: tcarbon = tons of carbon. SOURCE: Schipper, Ting et al. (1996).

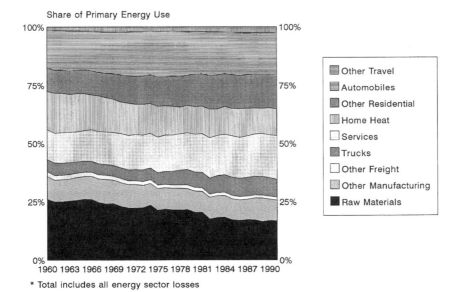

FIGURE 3-9 Evolution of energy use in the United States 1960-1992: From production to pleasure. SOURCE: Lawrence Berkley Laboratory calculations based on materials in Schipper, Ting et al. (1996).

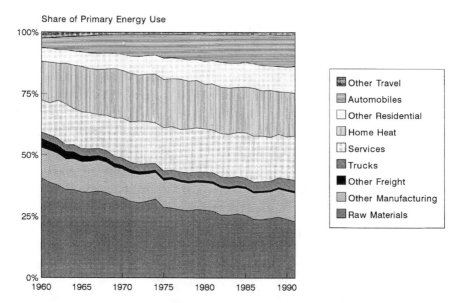

FIGURE 3-10 Evolution of energy use in the Federal Republic of Germany 1960-1992: From production to pleasure. SOURCE: National Energy Balances from the Federal Republic of Germany.

tance of households and personal transportation over the other sectors and means that energy uses are spreading from the largest users (factories) to the smallest users (households and the users of individual vehicles).

In personal transportation, activity increased principally in the more energy-intensive modes of cars and air, and overall energy intensities fell only in the United States. Significantly, the energy intensity of automobile travel in the United States, which fell from the mid-1970s until 1991, has risen since then. The real, on-road fuel intensity of new cars and household light trucks is no longer lower than that of the fleet of the same vehicles, signaling the end of an era that had seen the rapid decline in the energy intensities of new vehicles.

Scholl et al. (1996) analyzed the changes in travel emissions. Simplifying this work is the fact that travel depends almost wholly on oil products, for which emissions vary little from one fuel to the next. As a result, changes in emissions depend principally on changes in energy use. Figure 3-11 shows per capita emissions for travel by mode in 1973 and 1992 for Japan, the United States, and eight European countries aggregated. Because the differences among European countries, and changes in Europe over time were relatively uniform, these countries are aggregated to simplify the description.

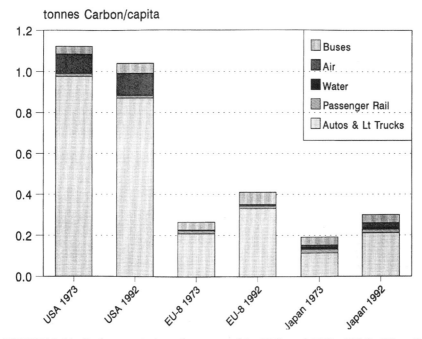

FIGURE 3-11 Carbon emissions from travel in 1973 and 1992. EU-8: West Germany, Denmark, Finland, Luxembourg, Norway, Sweden, France, and United Kingdom. NOTE: Lt: Light. SOURCE: Scholl, Schipper, and Kiang (1996).

Figure 3-12 shows the behavior of per capita aggregate emissions over time. The dips in the evolution of aggregate emissions were caused mainly by declines in activity during periods of recession and higher fuel prices. Note that the predominant trend was toward greater per capita emissions, however, principally from automobiles, except in the United States Not surprisingly, aggregate carbon intensity (Figure 3-13), the ratio of emissions to aggregate activity in passenger–kilometers, increased over time as well, decreasing only in the United States; in Denmark and Italy, there were marginal decreases relative to 1973.[1] In all but a few countries, it took more energy (and released more carbon) to transport a person one kilometer in 1992 than in 1973.

Figure 3-14 compares the different effects. In all countries/regions, activity (greater domestic travel per capita) boosted emissions, from a low of 31 percent in the United States to 65 percent in Japan and about 40

---

[1]In this analysis the unit of activity, passenger-kilometer, is calculated for automobiles as vehicle-kilometer times load factor, or people per car.

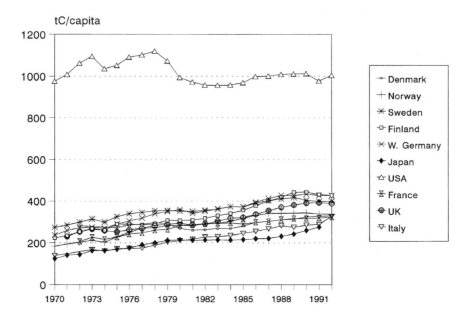

FIGURE 3-12   Travel carbon emissions per capita in 10 OECD countries.
NOTE: tC = tons of carbon. SOURCE: Scholl, Schipper, and Kiang (1996).

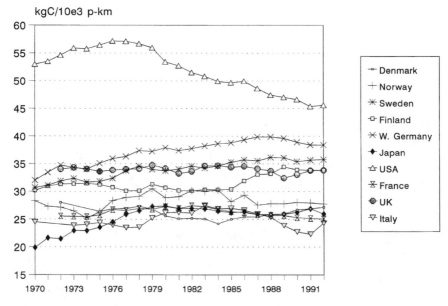

FIGURE 3-13   Aggregate carbon intensity of travel in 10 OECD countries.
SOURCE: Scholl, Schipper, and Kiang (1996).

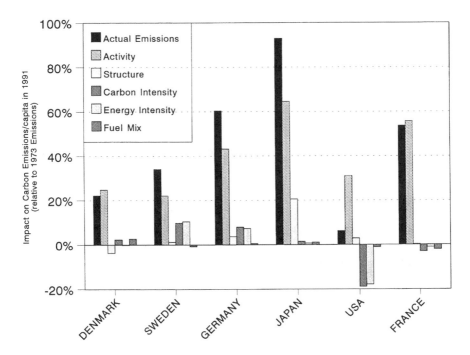

FIGURE 3-14 Impact of changes in components on carbon emissions from travel.
SOURCE: Scholl, Schipper, and Kiang (1996).

percent in European countries as a whole. Structural (modal) shifts to-ward cars and air travel increased emissions almost everywhere, above all in Japan, where automobiles only passed the 50 percent share of total travel in the late 1980s. Intensity changes reduced emissions by nearly 20 percent in the United States but had little effect elsewhere. This surprising result occurred because the load factors of automobile travel fell by 25 percent in every country, while the energy intensities of car use, in milli-joules per vehicle-kilometer, fell by less than 15 percent except in the United States, where they fell by 35 percent through 1991, at which time the decline stopped abruptly. Air travel energy intensity fell significantly in every country, but this change contributed marginally to total emis-sions, except in the United States. As a result of all of these changes combined, emissions per capita fell only in the United States, but in-creased in every other country, along with the share of carbon emissions from travel, as can be seen in Figure 3-11.

The high level of carbon emissions from travel in the United States relative to the other countries shown in Figure 3-11 is a function of the distances Americans travel. In particular, Americans travel 60-100 percent farther per capita by car than do Europeans. Surprisingly, however, an average car trip in either region is between 12.5 km and 15 km (Schipper, Gorham, and Figueroa, 1996). Thus, it is the frequency of car travel, not "distances" *per se*, that boosts Americans' travel.

ENERGY, EMISSIONS, AND LIFESTYLE

Examining variation in carbon emissions for the same activities among industrialized countries suggests which factors drive emissions and might lie behind potential for future restraint. I examine recent trends in automobile characteristics and use these as a window into the role in emissions of "lifestyle"—that is, the bundle of activities in which individuals engage (Schipper et al., 1989). A variety of indicators describe lifestyle: personal consumption expenditures, ownership of and access to energy-using consumer goods, time use, and distance traveled. Lifestyle "attributes" include the sociodemographic characteristics, such as age distribution or employment status of individuals and families. Lifestyle "choices" are activities that the population as a whole, sociodemographic subgroups, or individuals make—for example, choices on how much time to spend outside the home. These characteristics are not independent since families with small children may have to spend more time at home than families with no children. Some trends that increased energy use, such as increased numbers of women working (and driving to work) or smaller household size (which raised per capita area) can hardly be "faulted" for raising CO_2 emissions, but other trends, such as the purchase of larger homes or the gradual movement of households away from cities, must be considered conscious decisions, at least in part.

Income-driven lifestyle changes during the past decades have raised energy use for pleasure—i.e., for comfort and mobility. This is one way that lifestyle, as measured by the ownership and use of household equipment, travel, and visits to the service sector, continues to increase carbon emissions, even when those increases are less than proportional to income increases. Since the 1973 oil crisis, many energy uses have become less energy intensive, while lifestyles, continuing a long-term trend, have become more energy intensive. This is not a rebound effect: the largest energy savings and emissions reductions occurred in the household sector, where the largest declines in energy intensities occurred. During this period no energy savings occurred in travel, except in the United States.

The potential for technological change in transportation energy use (Schipper, Meyers et al., 1992; Schipper, 1993), is only being harvested slowly. Therefore at present the "structural" changes in these sectors drive energy use, as people change. Schipper et al. (1989) demonstrated that much of this change can be measured by following expenditures of money and time. As Gershuny and Jones (1987) demonstrated, most of us have more leisure time and spend increased amounts of that leisure away from home, consistent with what surveys of individual travel show (Schipper, Gorham, and Figueroa, 1996). As every person with a driver's license has at least one car, the characteristics of these devices and their overall use become more important in determining energy use, unless new energy-intensive technologies appear. Unless energy prices are extremely high, many of these choices will be made with little regard for energy prices. Although household energy uses appear saturated, no such trend is apparent for travel in the 1990s. In addition to increased car ownership, car characteristics and use have increasing importance to emissions.

Figure 3-15 shows an important indicator of automobile characteristics that affects emissions, new car weight, in the United States and some European countries (Schipper, 1995). Although the weight of a United States car fell significantly in the late 1970s, weight increased after the early 1980s; in contrast, weight of new cars in Europe appears to have increased continuously. When the rising share of light trucks is added to the United States figures, the rebound is more dramatic. Nevertheless Americans occupied considerably lighter new cars in 1993 than in 1973. Needless to add, the engine size, or horsepower, in Europe increased continuously, while the same parameters dropped and then slowly rebounded in the United States.

Figure 3-15 contrasts with the data on space-heating intensity, which fell 25-50 percent in OECD countries—so much so that even with modest increases in heated area, per capita space-heating energy use was lower in 1991 than in 1973 in almost all the countries we studied. But the analogous change in energy use for travel occurred only in the United States. To be sure, the ratio of fuel consumption to weight in new cars in virtually every OECD country has fallen continuously in all countries shown and, in fact, differs very little between these countries. Thus in a technological sense, cars are almost equally "efficient" in all countries, but their test fuel consumption still differs significantly because of weight, power, and other features. Because cars are heavier now than in 1980, actual fuel consumption per kilometer has fallen very little, except in the United States.

Figure 3-16 shows the use of cars in kilometer per capita per year; this reflects both the distances cars are driven and the number of cars per person. In the countries with the fewest cars (Finland, Britain, or Denmark), yearly usage per car is very high, accounting for the small range of car use shown within Europe. Australia lies slightly above the European

---

[2]Multiplying the values in this figure by the load factor, 1.5-1.7 people per car, gives the travel from cars. Only a small part of the United States-Europe gap in either cars or car use per capita is filled by much higher use of bus and rail in Europe. In fact, the car accounts for 80-85 percent of all travel in the European countries shown and accounts for 55 percent of travel even in Japan.

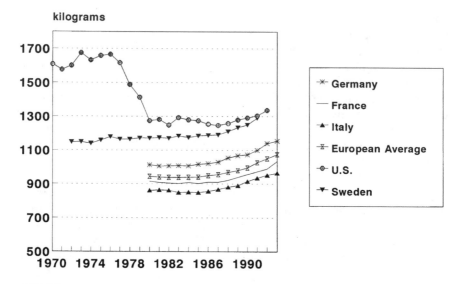

FIGURE 3-15 Average new car weight in Europe and the United States. Data excludes light trucks. Data from U.S. Department of Transportation, National Highway Traffic Safety Administration; Statistics Sweden; and European Association of Car Manufacturers.

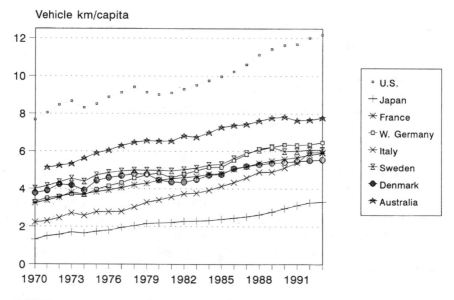

FIGURE 3-16 Per capita car use in 8 OECD countries. Data include household light trucks in the United States, Britain, and Denmark. Adapted from Lawrence Berkeley Laboratory.

countries, and Japan lies far below. The United States value appears to be moving away from the levels elsewhere.[2]

The U. S. data are no surprise in a country with fuel prices lying at one-third to one-fourth of the other countries in this study. Whatever the exact coupling between fuel prices and new car characteristics or use, the result of Americans' choices is three to four times the per capita carbon emissions from personal vehicles, mostly because of greater per capita driving but also because U. S. cars use 25-33 percent more fuel per km than those in Europe.

Unlike central heating, house area, and appliance ownership, which appear to be saturating (Schipper, in press), there is no evidence of such saturation for travel. And while new homes or appliances are significantly less energy intensive than those they replace or supplant (Schipper, Meyers, et al., 1992), the same cannot be said of automobiles, (although it is true for aircraft). Thus, although the household sector significantly reduced its CO_2 emissions in most countries studied, emissions from the travel sector were lower in 1991 than in 1973 only in the United States and started to rise after 1991 even there. Hence, travel is an area for concern.

In the future, only very slow growth can be expected in the structural factors that formerly pushed up household energy use because of slowing trends toward larger homes, smaller households, and ownership of major energy-using equipment. The travel sector, however, differs. Although gradual aging of the population may leave more of us home more often, and less mobile, roughly 30-40 percent of all Europeans of driving age still do not drive. These are mostly older people; among those in the 20-35 age group, car use is almost universal. Therefore, car use is expected to increase in Europe. Moreover, increases in driving in the United States and Europe are mainly to visit the service sector or for free time and holidays. Liberalization of shopping hours in Europe may encourage more evening and weekend car use. And the characteristics of new cars in Europe and the United States continue to evolve in ways that are more fuel intensive, offsetting much or all of the effort to reduce fuel use through technology. The high fuel prices in Europe will probably keep a permanent wedge in per capita fuel use between the United States and Europe. But in contrast to the situation in the household sector, all indicators of energy use and CO_2 emissions from travel now point up (Schipper, 1995).

This analysis shows that travel is emerging as the primary leader of growth in carbon emissions in the wealthy, industrialized countries. Lifestyle changes driven predominantly by higher incomes—particularly increased automobility—have consistently led to higher carbon emissions, and the trends in the travel sector show no signs of saturation. Because the energy intensity of travel is scarcely falling, coupling between lifestyles and emissions in the travel sector may cause difficulties for governments intent on restraining or even cutting emissions. It is critical to understand not only how efficiently energy is converted to energy services but also how the levels of services are growing.

ACKNOWLEDGMENT

The author is a Staff Senior Scientist with the International Energy Studies Group, Energy Analysis Program, Energy and Environment Division, Lawrence Berkeley Laboratory, and currently on leave to the International Energy Agency (I.E.A.), Paris. This work was initially supported by the U. S. Environmental Protection Agency and completed at the I.E.A. The opinions advanced are strictly those of the author.

REFERENCES

Gershuny, J., and S. Jones
  1987   *Time Use in Seven Countries, 1961 to 1984.* Bath, England: University of Bath.
Howarth, R.B., L. Schipper, and B. Andersson
  1993   Structure and intensity of energy use: Trends in five OECD nations. *Energy Journal* 14(2):27-45.
Schipper, L.
  1993   Energy Efficiency and Human Activity: Lessons from the Past, Importance for the Future. Report presented at the World Bank Development Conference. May 3-4, Washington, D.C.
  1995   Automobile use and energy consumption in OECD countries. *Annual Review of Energy and the Environment* 21. Palo Alto, Calif.: Annual Reviews Inc.
  in   People in the greenhouse: Indicators of carbon emissions from households and
  press   travel. In T. Dietz, ed., *Environmental Impacts of Consumption.*
Schipper, L., S. Bartlett, D. Hawk, and E. Vine
  1989   Linking lifestyles and energy use: A matter of time? *Annual Review of Energy* 14:273-320.
Schipper, L., R. Gorham, and M.J. Figueroa.
  1996   *People on the Move: Comparison of Travel Patterns in OECD Countries.* Paper prepared for the United States Department of Transportation. Berkeley, Calif.: Lawrence Berkeley Laboratory.
Schipper, L., R. Haas, and C. Sheinbaum
  1996   Recent trends in residential energy use in OECD countries and their impact on CO_2 emissions. *Journal of Mitigation and Adaptation to Global Changes* 1(2):167-196.
Schipper, L., F. Johnson, R. Howarth, B.G. Andersson, B.E. Andersson, and L. Price
  1993   *Energy Use in Sweden: An International Perspective* (LBL-33819). Berkeley, Calif.: Lawrence Berkeley Laboratory.
Schipper, L., S. Meyers, R. Howarth, and R. Steiner
  1992   *Energy Efficiency and Human Activity: Past Trends, Future Prospects.* Cambridge, England: Cambridge University Press.
Schipper, L., L. Scholl, and N. Kiang
  1996 CO_2 emissions from passenger transport. *Energy Policy* 24(1):17-30.

Schipper, L., L. Scholl, and L. Price
  in   Energy use and carbon emissions from freight in OECD countries. An analysis
 press of trends from 1973-1992. *Transport and the Environment.*
Schipper, L., M. Ting, M. Khrushch, F. Unander, P. Monahan, and W. Golove
  1996   *The Evolution of Carbon-Dioxide Emissions from Energy Use in Industrialized Coun-
        tries: An End-Use Analysis.* Berkeley, Calif.: Lawrence Berkeley Laboratory.
Scholl, L., L. Schipper, and N. Kiang
  1996 CO_2 emissions from passenger transport: A comparison of international trends
        from 1973-1992. *Energy Policy* 24(1):17-30.
Torvanger, A.
  1991   Manufacturing sector carbon dioxide emissions in nine OECD countries
        1973-1987. *Energy Economics* 13(2).

BIBLIOGRAPHY

Fergesson, M.
  1990   *Subsidized Pollution: Company Cars and the Greenhouse Effect.* London: Earth Re-
        sources Research.
Greening, L., W.B. Davis, and L. Schipper
  1996   Decomposition of Aggregate Carbon Intensity for Manufacturing: Comparison of
        Declining Trends from Ten OECD Countries for the Period 1971 to 1991. Unpub-
        lished paper submitted to *Energy Economics.*
Holdren, J.
  1992   Prologue: The transition to costlier energy. In L. Schipper and S. Meyers, with R.
        Howarth and R. Steiner, eds., *Energy Efficiency and Human Activity: Past Trends,
        Future Prospects.* Cambridge, England: Cambridge University Press.
Howarth, R., and R. Monahan
  1992   *Economics, Ethics, and Climate Policy* (LBL-33230). Berkeley, Calif.: Lawrence Ber-
        keley Laboratory.
Howarth, R.B., L. Schipper, P.A. Duerr, and S. Stroem
  1991   Manufacturing energy use in eight OECD countries. *Energy Economics*
        13(2):135-142.
Schipper, L., F. Johnson, R. Howarth, B.G. Andersson, B.E. Andersson, and L. Price
  1993   *Energy Use in Sweden : An International Perspective* (LBL-33819). 1992 Intergovern-
        mental Panel on Climate Change Supplement. Geneva, Switzerland: Intergov-
        ernmental Panel on Climate Change.
Schipper, L., and L. Price
  1994   Efficient energy use and well being: The Swedish example after 20 years. *Natural
        Resources Forum* 18(2):125-142.
Schipper, L., B. Richard, R. Howarth, B. Andersson, and L. Price
  1993   Energy use in Denmark: An international perspective. *Natural Resources Forum*
        17(2):83-103.
Schipper, L., B. Richard, R. Howarth, and E. Carlesarle
  1992   Energy intensity, sectoral activity, and structural change in the Norwegian
        economy. *Energy—The International Journal* 17(3):215-233.
Schipper, L., R. Steiner, M.J. Figueroa, and K. Dolan
  1993   Fuel prices and economy. *Transport Policy* 1(1):6-20.
Schipper, L., F. Unander, M. Khrushch, M. Ting, and L. Peraelae
  1996   *Energy Use in Ten OECD Countries: Long Term Trends through 1991.* Berkeley,
        Calif.: Lawrence Berkeley Laboratory.

Sheinbaum, C., and L. Schipper
    1993   Residential sector carbon dioxide emissions in OECD countries 1973-1989: A comparative analysis. Pp. 255-268 in *The Energy Efficiency Challenge for Europe: Proceedings of the ECEEE Summer Study, Vol. II.* Oslo, Norway: European Council for an Energy-Efficient Economy.

STRUCTURAL ECONOMICS: A STRATEGY FOR ANALYZING THE IMPLICATIONS OF CONSUMPTION

Faye Duchin

LIFESTYLES AND THE ENVIRONMENT

Interest in personal consumption is of long standing in economics, and many related aspects of household behavior have been addressed in all the social sciences. Consumption can be viewed as the motor propelling an economy in that producers will fabricate only the goods and services that consumers want to buy. Very recently, environmental concerns have reinvigorated social scientists' interest in consumption. Most environmental degradation can be traced to the extraction of fuels and other materials and their transformation to produce, both directly and indirectly, the goods and services valued by consumers. Clearly, changes surrounding consumption would alter, and could alleviate, pressures on the environment.

There are basically two ways in which such changes could be achieved. First, the technologies used to extract and transform materials could be improved in various ways. Second, consumption patterns could change. There are many efforts under way to develop technologies that are more efficient in their use of energy and materials and that generate less environmental damage than current practices. In this paper our concern is especially with consumption patterns. This paper describes a conceptual and methodological approach for situating consumption activities within a broad socioeconomic framework. It brings together various pieces of work that I have carried out over the past few years and fills in the missing pieces to make a relatively complete and coherent framework. A book-length manuscript that elaborates the major aspects of this approach has recently been completed (Duchin, 1996).

Economists are concerned with consumption by individuals, but there are two compelling reasons to think in somewhat broader terms. First, an individual's consumption behavior is tightly linked with his or her employment, in that earned income has to cover outlays for purchased goods and services. Consumption behavior is also related to other people's employment and consumption: if everyone stopped buying cars, auto workers (and, by a domino effect, many other workers) would soon be without jobs and income. Second, people live in households (including, of course, one-person households) that generally contain one or more paid workers. At least a portion of the income they earn is pooled, based on various kinds of negotiations, to pay for both common purchases and those of financially dependent individuals. A household's lifestyle refers to the jointly determined work and consumption practices of its members.

THE LOGIC OF STRUCTURAL ECONOMICS

There is a vast amount of literature indicating that consumer demand for specific goods levels off at higher incomes. However, prospects for overall saturation are far more ambiguous. Following J.S. Mill, a number of economists have expressed the view that once population levels off in affluent societies, other forms of satisfaction might be preferred to further purchases of goods and services (Mishan, 1967; Scitovsky, 1976; Hirsch, 1977), especially when people become aware of the environmental implications (Boulding, 1973; Daly, 1977). None of these authors, however, was able to provide an analytic framework for integrating these phenomena with other economic activities. The "new home economics" initiated by the work of Gary Becker, on the other hand, established the importance of the household as a decision-making unit within the analytic framework of neoclassical economics. Household decisions are portrayed as maximizing the household's "utility" subject to budget constraints; the treatment is analogous to that of business firms concerned only with maximizing their short-term profit (Becker, 1981).

Input-output economics provides a foundation for the description and analysis of household lifestyles that is both firmer and richer than neoclassical economics. However, this approach needs to be substantially extended in its coverage of both households and the physical environment. Structural Economics provides this extended framework.

Input-output economics describes the structure of an economy in terms of the interdependence among its different parts (Leontief, 1986). In the dynamic formulation, changes in structure result from technological changes, the accumulation of stocks of physical capital, and the depletion of stocks of resources. The framework consists of two simple but extremely flexible mathematical models—a model of physical interconnectedness and a corresponding representation of costs and prices—and a highly structured database.

In neoclassical economics, a money price needs to be associated with every variable, a network of parameters called elasticities govern automatic substitutions among inputs whenever prices change, economic actions are limited to the operations of competitive markets, and a solution requires that all markets are simultaneously in "equilibrium." The power of these assumptions is that they assure unique, optimal solutions to complex problems. However, the problems that are solved are arguably not

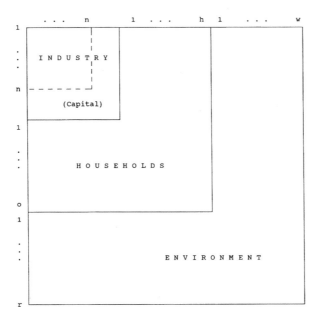

FIGURE 3-17 A structural table of an economy. NOTE: A structural table incorporates elements of an input-output table (*n* sectors), social accounting matrix (*o* occupations; *h* categories of household), and natural resource accounts (*r* resources, *w* categories of wastes). See text. SOURCE: Duchin (1995).

the most useful representation of actual situations. An input-output solution has the important advantages that it is not restricted to money values: substitutions of one technology for another are governed by scenarios rather than by formal mathematical expressions, and scenarios can reflect competitive behavior or behavior that is strategic, civic, or ethically motivated. Because many fewer kinds of assumptions are built into the formal framework, more burden falls on the development of scenarios and the interpretation of alternative outcomes.

Structural Economics situates the detailed inter-industry relationships within a broader social context of household activities, which in turn are entirely contained within the physical environment. Figure 3-17 shows the way in which a structural table extends an input-output table. The household and environmental portions of the table draw on social accounting (Stone, 1986; Keuning and de Ruijter, 1988) and natural resource accounting (Central Bureau of Statistics of Norway, 1992; de Haan et al., 1993; Lange and Duchin, 1994), respectively. Mathematical relationships that deal with these extensions in a realistic way have been developed in a number of recent studies.

A structural analysis starts by defining the questions that will be addressed and selecting or developing the mathematical model. A description of various input-output models can be found in Duchin (1988).

Then classifications for industries, households, and resources are established, and the "transactions" among them are quantified for one or more historical years. National Accounts can provide the bulk of this information.

The base year data are expressed as stocks or flows—e.g., tons of coal absorbed by the steel industry. The model is formulated in terms of both variables (representing the stocks and flows) and parameters; the latter quantify the relationships among variables. An example of a parameter is the tons of coal required, on average, to make a ton of steel in the United States in 1992, given the mix of technologies in use at that time and the relative importance of each. The mathematical equations specify the kinds of parameters that are required.

One or more scenarios are built for each of the questions to be explored. An example will demonstrate how a scenario translates the question into variables and parameters.

As part of an analysis of development strategies for Indonesia, we were asked what changes would be needed in agricultural technology for Indonesia to remain self-sufficient in food over the next several decades, while upgrading the quality of the diet for a growing population and being obliged to take some of the most fertile land out of food production (Duchin et al. 1993). This scenario required assumptions about changes in diet (i.e., consumption parameters) and in the yields of new agricultural technologies (i.e., parameters for the agricultural sectors). The computation would determine how much land would be required (i.e., endogenous variables) to support these assumptions.

Input-output case study methodology has been developed for structuring the data projections (Duchin and Lange, 1994). Case studies for Indonesia, focused on the use of land, water, and energy, were carried out for households, forestry, rice, other food crops, estate crops, livestock, pulp and paper, cement, chemicals, food processing, textiles and apparel, and basic iron and steel. The computations showed that even the most optimistic assumptions about the adoption of advanced agricultural technologies could not satisfy the land constraints and other requirements; food will need to be imported.

CATEGORIES OF HOUSEHOLDS

Standardized Industrial Classification (SIC) schemes for goods and services produced on farms and in factories and offices are in wide use. These classification schemes have made it possible to share data, compare across studies and across countries, and cumulate results. Standard Occupational Classifications also exist, although they are less widely used. Classification schemes for households are more fundamental than those for occupations but are at a much earlier stage of development.

Anthropologists and sociologists have provided detailed, qualitative description of specific categories of households; see Wilk and Lutzenheiser (Chapter 4, this volume) for recent developments. Economists have established classifications that cover the entire society, but they are usually in terms of income categories only. An exception is the work done within the social accounting framework. Most Social Accounting Matrices (SAMs) have been constructed to examine the distribution of income in developing countries.

A particularly detailed SAM (a flow table similar in structure to the industry and household portions of Figure 3-17) is the one for Indonesia, where households are classified according to urban or rural location, agricultural or nonagricultural nature of the work of the "head" of the household, and economic status, for a total of ten categories. This SAM also distinguishes four occupations and whether or not the workers are paid (Central Bureau of Statistics of Indonesia, 1990).

The most promising kind of household classification scheme is one developed for consumer research and marketing based on a direct examination of detailed data (by Jonathan Robbin; described in Weiss, 1988). Observing that people who share a "zip code" tend to have similar lifestyles, Robbin built a database about U.S. household practices in each of these small areas; he included detailed information from the Census of Households, automobile purchase lists, credit card information, voting records, social values from surveys carried out at the Stanford Research Institute, and a host of specialized, private surveys. Robbin discovered that 34 variables accounted for almost 90 percent of the variation among neighborhoods. Each zip code was rated on these variables and assigned to one of 40 clusters, for which Robbin created a descriptive name. The resulting classification, which is widely used by corporations and political candidates to customize their messages for specific markets, is shown in Table 3-5. Research scientists may well be able to improve on these categories for the kinds of purposes envisaged in this paper.

At the present time, my colleagues and I are designing classification schemes and building structural tables for several developing countries (Indonesia, the Dominican Republic, and Namibia) in collaboration with local researchers and the national statistical offices. The classification schemes are obviously very different from that shown for the United States in Table 3-5 but have been stimulated by its example. After this type of work has been done in several countries—with attention to using similar nomenclature for similar lifestyles—some categories are likely to emerge that are common to a variety of societies. The most important

TABLE 3-5 Household Classifications and Characteristics for the United States in 1987

ZQ	Cluster	Description
1	Blue Blood Estates	America's wealthiest neighborhoods includes suburban homes and one in ten millionaires
2	Money & Brains	Posh big-city enclaves of townhouses, condos and apartments
3	Furs & Station Wagons	New money in metropolitan bedroom suburbs
4	Urban Gold Coast	Upscale urban high-rise districts
5	Pools & Patios	Older, upper-middle-class, suburban communities
6	Two More Rungs	Comfortable multi-ethnic suburbs
7	Young Influentials	Yuppie, fringe-city condo and apartment developments
8	Young Suburbia	Child-rearing, outlying suburbs
9	God's Country	Upscale frontier boomtowns
10	Blue-Chip Blues	The wealthiest blue-collar suburbs
11	Bohemian Mix	Inner-city bohemian enclaves à la Greenwich Village
12	Levittown, USA	Aging, post-World War II tract subdivisions
13	Gray Power	Upper-middle-class retirement communities
14	Black Enterprise	Predominantly black, middle- and upper-middle-class neighborhoods
15	New Beginnings	Fringe-city areas of singles complexes, garden apartments and trim bungalows
16	Blue-Collar Nursery	Middle-class, child-rearing towns
17	New Homesteaders	Exurban boom towns of young, midscale families
18	New Melting Pot	New immigrant neighborhoods, primarily in the nation's port cities
19	Towns & Gowns	America's college towns
20	Rank & File	Older, blue-collar, industrial suburbs
21	Middle America	Midscale, midsize towns
22	Old Yankee Rows	Working-class rowhouse districts
23	Coalburg & Corntown	Small towns based on light industry and farming
24	Shotguns & Pickups	Crossroads villages serving the nation's lumber and breadbasket needs
25	Golden Ponds	Rustic cottage communities located near the coasts, in the mountains or alongside lakes
26	Agri-business	Small towns surrounded by large-scale farms and ranches
27	Emergent Minorities	Predominantly black, working-class, city neighborhoods
28	Single City Blues	Downscale urban singles districts
29	Mines & Mills	Struggling steeltowns and mining villages
30	Back-Country Folks	Remote, downscale, farm towns
31	Norma Rae-ville	Lower-middle-class milltowns and industrial suburbs, primarily in the South
32	Smalltown Downtown	Inner-city districts of small industrial cities
33	Grain Belt	The nation's most sparsely populated rural communities

% U.S. Household	Median Income	Home Value	% College Graduate
1.1	$70,307	$200,000+[a]	50.7
0.9	45,798	150,755	45.5
3.2	50,086	132,725	38.1
0.5	36,838	200,000+[a]	50.5
3.4	35,895	99,702	28.2
0.7	31,263	117,012	28.3
2.9	30,398	106,332	36.0
5.3	38,582	93,281	23.8
2.7	36,728	99,418	25.8
6.0	32,218	72,563	13.1
1.1	21,916	110,669	38.8
3.1	28,742	70,728	15.7
2.9	25,259	83,630	18.3
0.8	33,149	68,713	16.0
4.3	24,847	75,354	19.3
2.2	30,077	67,281	10.2
4.2	25,909	67,221	15.9
0.9	22,142	113,616	19.1
1.2	17,862	60,891	27.5
1.4	26,283	59,363	9.2
3.2	24,431	55,605	10.7
1.6	24,808	76,406	11.0
2.0	23,994	51,604	10.4
1.9	24,291	53,222	9.1
5.2	20,140	51,537	12.8
2.1	21,363	49,012	11.5
1.7	22,029	45,187	10.7
3.3	17,926	62,351	18.6
2.8	21,537	46,325	8.7
3.4	19,843	41,030	8.1
2.3	18,559	36,556	9.6
2.5	17,206	42,225	10.0
1.3	21,698	45,852	8.4

continued on next page

TABLE 3-5 Continued

ZQ	Cluster	Description
34	Heavy Industry	Lower-working-class districts in the nation's older industrial cities
35	Share Croppers	Primarily southern hamlets devoted to farming and light industry
36	Downtown Dixie-Style	Aging, predominantly black neighborhoods, typically in southern cities
37	Hispanic Mix	America's Hispanic barrios
38	Tobacco Roads	Predominantly black farm communities throughout the South
39	Hard Scrabble	The nation's poorest rural settlements
40	Public Assistance	America's inner-city ghettos

National Median

NOTE: The source document does not report the year for which the data apply or the price unit. The household percentages are based on 1987 data, but the values appear to be for 1986 in current prices. The table shows a median household income of $24,269; this compares with figures of $23,618 for 1985 and $24,897 for 1986, according to the U.S. Bureau of the Census, *Statistical Abstract of the United States (1994)*, Table No. 707. The ZQ (zip qual-

lifestyle changes in the developing countries surround those households whose work is unregistered and untaxed, and who are largely not reached by social services. Their ways of life are being rapidly altered by urbanization and industrialization. The objective of scenario analysis in this context is to anticipate the nature and magnitude of these changes in terms, for example, of the future demand for education, health care, sanitary facilities, or small loans.

REFERENCES

Becker, G.S.
    1981   *A Treatise on the Family*. Cambridge, Mass: Harvard University Press.
Boulding, K.
    1973   The economics of the coming spaceship earth. Pp. 253-263 in H.E. Daly, ed.,
            Economics, Ecology, Ethics. San Francisco: W.H. Freeman.
Central Bureau of Statistics of Indonesia
    1990   *Social Accounting Matrix for Indonesia, 1985*. Jakarta, Indonesia.
Central Bureau of Statistics of Norway
    1992   *Natural Resources and the Environment 1991*. Oslo, Norway.
Daly, H.
    1977   *Steady-State Economics*. San Francisco: W.H. Freeman.

% U.S. Household	Median Income	Home Value	% College Graduate
2.8	18,325	39,537	6.5
4.0	16,854	33,917	7.1
3.4	15,204	35,301	10.7
1.9	16,270	49,533	6.8
1.2	13,227	27,143	7.3
1.5	12,874	27,651	6.5
3.1	10,804	28,340	6.3
	$24,269	$64,182	16.2

ity) index, based on income, home value, education, and occupational status, measures socioeconomic rank.

[a]The upper census limit for home values is $200,000+; the figures for Blue Blood Estates and Urban Gold Coast are estimates.

SOURCE: Duchin (1995), based on Weiss (1988) pp. 4, 5, 12, 13.

de Haan, M., S. Keuning, and P. Bosch
   1993   *Integrating Indicators in a National Accounting Matrix Including Environmental Accounts.* Netherlands Central Bureau of Statistics, No. NA-060.
Duchin, F.
   1988   Analyzing structural change in the economy. In M. Ciaschini, ed., *Input-Output Analysis: Current Developments.* London: Chapman and Hall.
   1995   Global Scenarios about Lifestyle and Technology. Paper prepared for the Sustainable Future of the Global System conference, United Nations University, Tokyo, Japan.
   in   Household Lifestyles: The Social Dimension of Structural Economics. Paper
   press   prepared for the United Nations University, Tokyo, Japan.
Duchin, F., C. Hamilton, and G. Lange
   1993   Environment and Development in Indonesia: An Input-Output Analysis of Natural Resource Issues. Final report for Indonesian Ministry of Planning. U.S. Agency for International Development and Canadian International Development Agency.
Duchin, F., and G. Lange
   1994   *The Future of the Environment: Ecological Economics and Technological Change.* New York: Oxford University Press.
Hirsch, F.
   1977   *Social Limits to Growth.* London: Routledge and Kegan Paul.
Keuning, S., and W. De Ruijter
   1988   Guidelines to the construction of a social accounting matrix. *Review of Income and Wealth.* Series 34. 1(March): 71-100.

Lange, G., and F. Duchin
    1994   *Integrated Environmental-Economic Accounting*. Natural Resource Accounts, and Natural Resource Management in Africa. Washington, D.C: Winrock International Environmental Alliance.

Leontief, W.
    1986   *Input-Output Economics*, 2nd ed. New York: Oxford University Press.

Mishan, E.J.
    1967   *The Costs of Economic Growth*. New York: Penguin Books.

Scitovsky, T.
    1976   *The Joyless Economy*. New York: Oxford University Press.

Stone, R.
    1986   Social accounting: The state of play. *Scandinavian Journal of Economics:* 453-472.

U.S. Bureau of the Census
    1994   *Statistical Abstract of the United States*. Table No. 707. Washington D.C.: U.S. Department of Commerce.

Weiss, M.J.
    1988   *The Clustering of America*. New York: Harper and Row.

4

Examining the Driving Forces

INTRODUCTION

Any informed effort to address the environmental impacts of consumption must begin with an understanding of what causes, or drives, environmentally important consumption activities. Economics has made major contributions to understanding consumption by considering prices, budgetary constraints on choice, the costs of information about alternative actions, the ability to externalize costs, and so forth. It has also emphasized the fact that consumption and production are elements of a dynamic system in which all the elements respond together to external events, so that the environmental impacts of consumption are intimately tied to those of production. These economic insights are essential for understanding the dynamics of consumption.

Understanding environmentally significant consumption also requires the use of concepts not normally included in economic analyses. For example, economics normally treats preferences as exogenous to analyses, presuming that during the time frame of interest, preferences are constant. This assumption may not be reasonable when the analysis concerns human responses to long-delayed environmental changes such as in climate or the ozone layer, because the responses may occur over a period of several decades. In conducting such analyses, it is important to examine the possibility of change in preferences for at least two reasons. One is that preferences often change on time scales of a human generation or longer: it has been argued, for instance, that cohorts raised in an environment of affluence have different values and personal and policy preferences from cohorts raised with scarcity (Inglehart, 1990). Another reason to treat preferences as endogenous in environmental research is that information about impending environmental threats may be the sort of stimulus that causes people to reconsider their preferences. Thus, for studies of consumption and the environment, it may be important to consider processes such as preference construction and cultural change that may mediate the effects of standard economic variables.

This chapter presents five brief reports from the workshop that examine driving forces of consumption other than those usually addressed in economic analyses or that consider the relationships between economic forces and other factors. As in Chapter 3, the reports raise some intriguing questions for research and, through their bibliographies, direct readers to broader related literatures.

Loren Lutzenhiser's research examines residential energy use in northern California. The analysis includes several physical and economic explanatory variables typically used in this field, such as climate, dwelling size and type, appliance ownership, and household size and income. It also includes some factors not usually included in energy analysis, such as race, ethnicity, and cultural assimilation among relatively recent immigrant populations. Lutzenhiser finds that after taking climate, housing characteristics, household technology, and income into account, Hispanic and Asian households use less energy than whites, that African-Americans use more, and that the immigrant populations studied move toward the white American pattern as a function of acculturation, reflected by the language spoken in the household. The findings help address the question of how adoption of an American lifestyle alters household energy use and, through it, affects the environment. They suggest that immigrants may adopt patterns of energy use that are typically American over a generation or two.

The report by Thomas Dietz and Eugene Rosa uses a multivariate analytic approach to examine the effects of two driving forces on an indicator of environmental impact and reveal variations that can be attributed to other forces. They analyze national-level data on carbon dioxide emissions and estimate the effects of levels of population and affluence. They find a nearly linear effect of population and an effect of affluence (GNP per capita) that reaches a maximum at about U.S. $10,000 and then begins to decline. When the effects of population and affluence are estimated by regression, the residual variations cover more than a 20-fold range, probably attributable to national differences in technology, institutions, and other factors. Further study of the residual variation is one approach to clarifying the importance of driving forces other than population size and economic activity.

Eugene Rosa's analysis distinguishes measures of gross economic activity from other indicators of material well-being, analyzes the relationships among these other measures, and considers how they relate to an indicator of environmental change. He identifies four distinct composite indicators of nonfinancial material well-being and finds that all the affluent economies studied continued to change montonically on these indicators through 1985, even though in some of them the oil-market events of the 1970s altered the direction of the trend in carbon emissions per capita. He concludes that the transitions in these countries reflect a shift to more service-based, postmodern economies, in which both gross domestic product and nonmonetary indicators of welfare became less tightly coupled to carbon emissions during that period. Rosa suggests that further reductions in resource consumption can be made with only limited impacts on welfare.

Richard Wilk's report considers the hypothesis that Western styles of consumption have global environmental effects because people in developing countries emulate this consumption. Some scholars have inferred that exposure to Western cultural influence, through such media as exported films and television programs, drives consumption patterns in developing countries where per capita income is increasing. Such emulation matters for environmental policy because if increasingly affluent populations in developing countries mimic affluent Western lifestyles, there would be very serious global environmental impacts. If they adopt less resource-intensive and polluting styles of affluence, however, there might be great environmental benefits. Wilk identifies several indicators of emulation, notes their serious limitations to date, and presents his tentative reading of the data: that Western-style consumption is not a single package that consumers everywhere accept but, rather, that people of increased means in developing countries may pursue a variety of consumption aspirations and lifestyles. Despite Western mass media penetration of developing countries, Wilk finds only weak evidence that American middle-class consumer aspirations have been uniformly accepted. The question of whether there is emulation of the most environmentally damaging types of Western-style consumption has barely begun to be examined.

The report by Willett Kempton and Christopher Payne considers major social transformations in human history and prehistory as influences on both consumption (of energy and materials) and quality of life. They suggest that in the sweep of human history, increases in consumption have been driven by grand transformations of social structures but that these transformations, at least on some indicators such as health and leisure time, have not been associated with monotonic increases in quality of life. This analysis raises the question of whether forms of social organization might be adopted that provide an acceptable quality of life at much lower levels of materials and energy consumption than now exist in the industrialized world.

These five reports suggest some of the possibilities for investigating the effects of social and cultural phenomena on environmentally relevant consumption, either independently of standard economic variables or in interaction with them. There are, of course, many other such investigations that could be conducted. In Chapter 5, we discuss some strategies for setting priorities among the vast range of possible research questions linking consumption and the environment.

REFERENCE

Inglehart, R.
  1990   *Culture Shift in Advanced Industrial Society.* Princeton, N.J.: Princeton University Press.

SOCIAL STRUCTURE, CULTURE, AND TECHNOLOGY: MODELING THE DRIVING FORCES OF HOUSEHOLD ENERGY CONSUMPTION

Loren Lutzenhiser

This paper reviews some alternative conceptions of household energy consumption and uses an analysis of patterns of energy use in a California sample to demonstrate the joint influence of social status, ethnicity, and material culture in the structuring of energy flows. These findings suggest that conventional models of consumption obscure the workings of sociotechnical systems, seriously limiting our ability to understand the dynamics of energy consumption. Implications for scientific research and policy modeling, cross-cultural analysis, and environmental justice are also considered.

SOCIAL CONSUMPTION AND ENVIRONMENTAL CHANGE

Despite the current hiatus in public and policy concern about energy, the environmental impacts of energy use are increasingly clear. In fact, efforts to empirically examine, theorize, and model the dynamics and consequences of societal energy use have been pursued for more than 20 years. But understanding energy consumption is a far from straightforward matter. Although it is fairly obvious that energy flows are produced and shaped by human action, this consumption only occurs via a complex of fuel flows, energy-conversion technologies, and loosely coupled economic marketing/regulatory systems. And, as energy is consumed at many different end-use sites, and under fluctuating environmental conditions, the flow is determined by a fairly complex interplay of sociocultural, geographic, technological, and institutional factors. Because we lack an overarching interdisciplinary approach to such human-environment interactions (Stern 1993), efforts to understand this system have too often been narrowly focused—resulting in partial views of the system and its environmental impacts.

The social sciences have produced a fairly rich body of work on the role of energy and energy technology in society (e.g., see Cottrell, 1955; Mazur and Rosa, 1974; White, 1975; Adams, 1975; Buttel, 1979; Duncan, 1978; Olsen, 1991; Humphrey and Buttel, 1982; see Rosa et al., 1988, and Lutzenhiser, 1994, for reviews). A large literature also focuses on the connections between social status and consumption in general (Veblen, 1899; Weber, 1978; Lynes, 1955; Packard, 1959; Douglas and Isherwood, 1979; Mason, 1981; Mukerji, 1983; Fussell, 1983; Bourdieu, 1984; Forty,

1986; Miller, 1987; McCracken, 1988; Otnes, 1988; Saunders, 1990; Warde, 1990; Burrows and Marsh, 1992), including the construction of status via the stylized consumption of food, clothing, music, language, automobiles, housing, and appliances (Ewen, 1976, 1988; Cowan, 1989; Featherstone, 1990, 1991; Gartman, 1991). We know much less about how technology-shaping processes work in the institutional environment— e.g., how devices and machines come to have the energetic and stylistic features that they do, and how producers and consumers interact in the negotiation of design (Bijker et al., 1989; Bijker and Law, 1992). And, with a few exceptions (e.g., Uusitalo, 1983), little attention has been paid until quite recently to the linkages between culture, consumption, and the natural environment [see Durning, 1992; Brown, 1989; Schnaiberg's (1991) critique of Brown; Lutzenhiser and Hackett, 1993].

The interdisciplinary literature concerned directly with the consumption of energy suggests, however, that social structure and cultural practice are indeed central to the structuring of energy consumption (Lutzenhiser, 1992a; Lutzenhiser and Hackett, 1993), for significant energy use differences are observed between income groups (Newman and Day, 1975; Lacy, 1985; Skumatz, 1988), across life cycle stages (Frey and LaBay, 1983), and among ethnic subcultures (Kohno, 1984; Throgmorton and Bernard, 1986; Hackett and Lutzenhiser, 1991). Conservation behavior is also quite socially variable (Heberlein and Warriner, 1982; Dillman et al., 1983; Stern et al., 1986; Schwartz and True, 1990; Hackett and Lutzenhiser, 1991). Unfortunately, many of these studies have overlooked important housing and technology differences between social groups— "technical" variables that influence consumption.

Conventional energy policy models do little better, however, often glossing over the sociocultural aspects of energy use and choosing instead to treat "stocks" of buildings and equipment as the molar elements of a thoroughly technical analysis. Although the weaknesses in such approaches are well known (Stern, 1984, 1986; Stern and Aronson, 1984; Archer et al., 1984; Cramer et al., 1985; Baumgartner and Midtunn, 1987; Lutzenhiser, 1992b, 1993, 1994), these models continue to dominate policy discourse and the generation of energy system inputs for environmental systems modeling.

This disconnect between approaches focused exclusively on either the "social" or the "technical" aspects of energy consumption can, in fact, be overcome through a fairly straightforward synthesis. The following empirical case shows that consumption can, at once, be seen as shaped by the social allocation of buildings and equipment with energetic characteristics and by the cultural expression of energy-using behaviors.

A SOCIOTECHNICAL ANALYSIS OF
HOUSEHOLD ENERGY CONSUMPTION

The data used in the analysis are from a major survey of housing, energy use, and household technology in northern California (California Energy Commission, 1986). Rather than consumption being homogenous—as many of the simplest conventional models assume—these data show considerable variation in energy consumption across sample households (Figure 4-1), with distinct differences in consumption by subgroups defined on the basis of both "social" (e.g., life cycle stage, wealth, ethnicity) and " technical " (e.g., age, type, and size of housing and appliances) characteristics (Table 4-1). Because the social and technical aspects of consumption are correlated in these sorts of data, a series of multivariate models were estimated, one of which is reported in Table 4-2. This sociotechnical model offers a good fit to the data and suggests that both the behavior of social groups and their material conditions contribute to the structuring of consumption in a variety of ways. A second-stage analysis using regression estimates and subgroup characteristics shows that various combinations of behavior, housing, and technology are responsible for shaping consumption quite differently across social groups (Table 4-3). Rates of input energy waste and carbon dioxide pollution

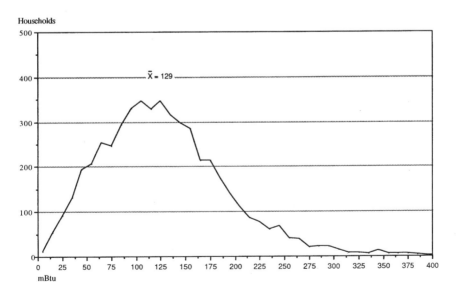

FIGURE 4-1 Annual Household Energy Consumption. Data from California Energy Commission (1986).

TABLE 4-1 Physical and Social Variation in Energy Consumption, Northern California Households

	Mean consumption (mBtu)	SD	n	Cases (%)
Entire sample	129	(68)	4127	100
Building size (sq ft)				
< 400	69	(43)	160	4
400-599	71	(51)	311	8
600-999	83	(40)	877	21
1,000-1,499	123	(49)	1216	29
1,500-1,999	150	(52)	883	21
2,000-2,699	182	(63)	486	12
2,700-3,499	214	(91)	139	3
> 3,500	250	(149)	55	1
Housing type				
Single family detached	156	(65)	2357	57
Multi-family	86	(48)	1770	43
Year dwelling built				
1979-84	112	(63)	651	16
1970-78	128	(69)	909	22
1960-69	138	(70)	771	19
1950-59	134	(64)	730	18
1940-49	124	(63)	376	9
pre-1940	116	(72)	690	17
Number of persons in household				
1	78	(45)	796	21
2	127	(63)	1454	39
3	144	(63)	630	17
4	160	(67)	511	14
5	165	(66)	197	5
> 6	175	(94)	118	3
Annual income (1986 dollars)				
< $10,000	93	(48)	635	15
$10,000-19,999	106	(55)	771	19
$20,000-29,999	118	(58)	772	19
$30,000-39,999	131	(60)	666	16
$40,000-49,999	139	(62)	453	11
$50,000-75,000	157	(71)	546	13
> $75,000	187	(105)	284	7

TABLE 4-1 Continued

	Mean consumption (mBtu)	SD	n	Cases (%)
Race/ethnicity and language spoken at home				
White	130	(70)	3349	83
Black	119	(62)	154	4
Hispanic	117	(54)	143	4
Hispanic (Spanish)	95	(48)	124	3
Asian	110	(66)	138	3
Asian (other)	106	(54)	130	3

NOTE: SD = Standard deviation.

were also found to be socially variable (Table 4-4). When conventional approaches focus on "typical" households and amorphous stocks of housing, they fail to take these sorts of social variations in consumption into account.

UNDERSTANDING THE SOCIAL NATURE OF MATERIALS SYSTEMS

A fundamental reorientation of theory is needed. The material environment can usefully be seen as an evolving social system in which social status (accomplished through status-graded buildings, equipment, and behavior) is a primary determinate of energy consumption, waste, and pollution. In a system of status-graded lifestyles, volumes of energy flow provide rough measures of social standing—the poor being excluded from all but modest forms of consumption, the middle classes sustained by consumption centered largely in housing and technologies, and the wealthy empowered in a variety of ways by high levels of energy flow. Rather than the amorphous housing stock assumed in energy analysis, occupied structures actually compose an ordered artificial environment, elaborated over time, its present form reflecting the realities of topography and climate; historical access to materials; the costs of land, labor, and energy availability (a mirror of past political economy); as well as past technical knowledge and cultural preference.

The built environment is a physical accretion of the products of sociotechnical change—literally embodying historical social arrangements (e.g., family size and class structure) in built forms—forms to which present occupants must behaviorally adapt. In treating buildings and

TABLE 4-2 Regression of Annual Energy Consumption on Social, Housing, Technology, and Environmental Variables

	Energy Consumption (mBtu)		
	b	SE	p
Household characteristics			
N of children <18 yr (0–8)	7.5	(0.9)	a
N of adults (0–11)	6.8	(0.9)	a
African-American	14.4	(4.6)	a
Hispanic-English (spoken at home)	−2.8	(4.6)	
Hispanic-Spanish	−12.3	(5.1)	b
Asian-English	−12.7	(4.9)	b
Asian-other	−26.7	(5.1)	a
<$15,000	4.4	(2.2)	c
$15,000-34,999	−1.4	(2.4)	
>$50,000	15.0	(2.5)	a
Person(s) at home during day	7.0	(1.8)	a
Housing characteristics			
Dwelling size (1,000 sq ft)	27.6	(1.0)	a
Multi-family unit (attached)	−17.8	(2.3)	a
Built after 1979 (energy building codes)	−9.7	(2.6)	a
Building energy efficiency scale (1-6)	−1.7	(0.6)	b
Air conditioning	13.3	(2.4)	a
Solar water heating	−2.3	(4.9)	
Household technology			
Clothes washer	7.9	(3.6)	c
Clothes dryer	12.5	(3.3)	a
Dishwasher	10.3	(2.0)	a
Frost-free refrigerator	9.4	(2.2)	a
2+ refrigerators	20.2	(2.4)	a
Freezer	12.4	(2.0)	a
Other appliances (0-7)[d]	2.4	(0.7)	a
Pool, hot tub, or spa	36.1	(3.5)	a
Environment			
CEC1	−3.5	(4.0)	
CEC2	−8.0	(3.0)	b
CEC4	−0.9	(2.9)	
CEC5	−3.7	(3.3)	
(Intercept)	22.0	(5.5)	

NOTE: b = slope of regression line, signifying mBtu consumed per unit of the independent variable; SE = standard error of b; CEC = dummy variable signifying climatic regions; mBtu = million British thermal units.

[a]$p < .001$; [b]$p < .01$; [c]$p < .05$; [d]Color TV, computer, stereo, black and white television set, microwave, video, humidifier.

technologies as the primary "actors" in society-environment relations, conventional models claim a fictive autonomy for physical objects—divorcing them from the social structures and cultural processes within which they are embedded and from which they necessarily derive. When used to inform policy, these approaches also import biases—masking important social differences in material conditions and behavior.

IMPLICATIONS FOR RESEARCH AND POLICY ANALYSIS

A number of basic scientific and policy research implications follow from these findings. A considerable amount of fruitful work might be done, for example, in examining empirical patterns of consumption and disaggregating their sources through time across the United States. Linkages between energy-use patterns and the patterned consumption of other goods and services (automobiles, food, entertainment, travel, etc.) might also be explored. And, the influences of a wider range of lifestyle orientations than can be captured by simple demographic categories should also be examined.

Studies of consumption that compare U.S. patterns with those found elsewhere in the industrialized world would also be useful. These studies could extend to the consumption of energy "embodied" in goods and services (a significant fraction of overall consumption). It would also be valuable to inventory and compare other resource flows (water, food, paper, metal, plastic, packaging) and waste flows (garbage, sewage, atmospheric emissions). And, a good deal of attention is overdue to the social patterning of transportation and gasoline consumption—a significant source of energy demand and environmental pollution.

Policy implications also follow from the social variation in consumption, the persistence of some low-energy-use cultural patterns in the midst of affluence, and the failure of conventional models to capture these variations. Policy-oriented research might focus on how conventional modeling systems operate and persist, and how cultural and institutional factors might be introduced to energy-policy modeling. Ethnographic work on cultural differences in consumption could shed light on the roots of persistence of low consumption levels and might suggest how durable and long-lived those patterns might be. Studies of "social traps" in housing and technology—both for the poor and the relatively more affluent—might reveal policy openings and long-term problems with consumption rooted in settlement patterns and social institutions (e.g, property-ownership conventions, taxation, inheritance, and lending systems). The implications for equity and community that follow from a more social model of built environment and energy use are also significant in a more populous, competitive, and highly engineered future. The growth of consumption

TABLE 4-3 Mixture of Consumption Sources: Model Household Types

White $ Older single

Persons	9	.12
Lifestyle	22	.29
Building	19	.25
Technology	26	.34
TOTAL	76	1.00

White $$ Young/small family

Persons	34	.25
Lifestyle	18	.13
Building	37	.27
Technology	49	.35
TOTAL	138	1.00

White $$ Older couple

Persons	19	.14
Lifestyle	18	.13
Building	45	.34
Technology	51	.38
TOTAL	133	1.00

Black $ Older single

Persons	9	.12
Lifestyle	35	.47
Building	10	.13
Technology	20	.27
TOTAL	74	1.00

Black $ Young/small family

Persons	34	.29
Lifestyle	36	.32
Building	16	.14
Technology	29	.25
TOTAL	115	1.00

Black $$ Young/small family

Persons	30	.26
Lifestyle	31	.27
Building	27	.23
Technology	28	.24
TOTAL	115	1.00

Hispanic(Spanish)$ Young/small family

Persons	38	.44
Lifestyle	11	.13
Building	16	.18
Technology	22	.25
TOTAL	86	1.00

Hispanic(English) $ Young/small family

Persons	38	.37
Lifestyle	20	.19
Building	20	.19
Technology	25	.24
TOTAL	103	1.00

Hispanic(Spanish) $$ Young/small family

Persons	36	.35
Lifestyle	6	.06
Building	28	.27
Technology	33	.32
TOTAL	103	1.00

White $$$ Young couple

Persons	19	.15
Lifestyle	18	.14
Building	41	.31
Technology	53	.40
TOTAL	130	1.00

White $$$ Young/small family

Persons	36	.23
Lifestyle	18	.11
Building	46	.29
Technology	57	.36
TOTAL	157	1.00

White $$$$ Young couple

Persons	19	.12
Lifestyle	32	.21
Building	47	.30
Technology	58	.37
TOTAL	156	1.00

Hispanic(English) $$
Young/small family

Persons	37	.32
Lifestyle	14	.12
Building	31	.26
Technology	34	.30
TOTAL	116	1.00

Asian(English) $$ Young couple

Persons	18	.25
Lifestyle	6	.08
Building	24	.32
Technology	27	.36
TOTAL	74	1.00

White $$$$ Young/small family

Persons	37	.19
Lifestyle	33	.17
Building	58	.30
Technology	66	.34
TOTAL	193	1.00

Asian(Other) $$ Older/extended family

Persons	44	.37
Lifestyle	0	.00
Building	35	.29
Technology	40	.34
TOTAL	119	1.00

Asian(English) $$$ Young/small family

Persons	37	.26
Lifestyle	6	.04
Building	45	.31
Technology	57	.39
TOTAL	145	1.00

White $$$$ Older/extended family

Persons	39	.20
Lifestyle	33	.17
Building	59	.30
Technology	67	.34
TOTAL	197	1.00

NOTE: Units are million British thermal units and fractions of total.
$ = low income; $$ = lower middle income; $$$ = upper middle income; $$$$ = high income.

TABLE 4-4 Race/Ethnicity and Class Patterns of Energy Intensity, Waste, and Pollution

	Energy Intensity			
	Living space (sq ft/person)	Total energy (mBtu)	Energy per capita (mBtu)	Energy per sq ft (mBtu)
Entire sample	650	126	58	100
White	685	130	62	99
< $15,000	648	99	61	105
$15,000-50,000	668	126	60	99
> $50,000	771	173	68	95
Black	555	119	55	125
< $15,000	550	103	59	131
$15,000-50,000	547	126	52	123
> $50,000	615	164	56	101
Hispanic (English)	433	117	44	116
< $15,000	361	99	44	136
$15.000-50,000	444	118	42	107
> $50,000	532	150	50	103
Hispanic (Spanish)	366	95	33	107
< $15,000	361	93	35	113
$15,000-50,000	364	98	29	99
> $50,000	569	110	49	79
Asian (English)	596	110	42	86
< $15,000	448	69	31	84
$15,000-50,000	579	109	39	87
> $50,000	727	136	56	85
China, Japan, S. Asia	422	106	30	92
< $15,000	448	87	30	99
$15,000-50,000	372	108	30	95
> $50,000	521	123	29	74

NOTE: mBtu = million British thermal units.
 [a]tons of carbon.
 [b]pounds of carbon.

Waste and Pollution				
Waste energy (mBtu)	CO_2 (tons)[a]	Waste CO_2 (tons)[a]	Waste energy per capita (mBtu)	CO_2 per capita (lbs)[b]
68	2.8	1.5	27	2,232
70	2.9	1.5	29	2,383
54	2.3	1.2	28	2,379
68	2.8	1.5	27	2,208
93	3.8	2.0	33	2,693
65	2.8	1.4	26	2,216
57	2.4	1.2	27	2,305
69	2.9	1.5	25	2,086
86	3.6	1.9	26	2,194
62	2.5	1.3	20	1,619
51	2.2	1.1	16	1,387
63	2.5	1.3	20	1,613
79	3.1	1.7	25	1,956
50	2.1	1.1	14	1,211
49	2.1	1.1	14	1,235
52	2.1	1.1	14	1,116
54	2.6	1.3	20	1,889
62	2.5	1.3	23	1,859
41	1.7	0.9	16	1,384
61	2.4	1.3	22	1,729
74	3.1	1.6	27	2,304
60	2.4	1.3	16	1,253
48	1.9	1.0	13	1,011
60	2.4	1.3	16	1,274
72	2.9	1.5	18	1,455

and the decline of ecosystem resources—even with the benefits of advanced environmental technologies—are likely to pose serious problems for even the most prosperous peoples and places in the industrialized world.

REFERENCES

Adams, R.
  1975   *Energy and Structure: A Theory of Social Power.* Austin, Tex.: University of Texas Press.
Archer, D., M. Costavnzo, B. Iritani, T.F. Pettigrew, I. Walker, L.T. White
  1984   Energy conservation and public policy: The mediation of individual behavior. Pp. 69-92 in W. Kempton, M. Neiman, eds., *Energy Efficiency: Perspectives on Individual Behavior.* Washington, D.C: ACEEE Press.
Baumgartner, T., and A. Midtunn, eds.
  1987   *The Politics of Energy Forecasting: A Comparative Study of Energy Forecasting in Western Europe and North America.* Oxford, England: Clarendon Press.
Bijker, W., T.P. Hughes, and T. Pinch, eds.
  1989   *The Social Construction of Technological Systems.* Cambridge, Mass.: M.I.T. Press.
Bijker, W., and J. Law, eds.
  1992   *Shaping Technology/Building Society: Studies in Sociotechnical Change.* Cambridge, Mass.: M.I.T. Press.
Bourdieu, P.
  1984   *Distinction: A Social Critique of the Judgment of Taste.* Cambridge, Mass.: Harvard University Press.
Brown, L.
  1989   *State of the World.* New York: W.W. Norton.
Burrows, R., and C. Marsh, eds.
  1992   *Consumption and Class: Divisions and Change.* London: Macmillan.
Buttel, F.H.
  1979   Social welfare and energy intensity: A comparative analysis of the developed market economies. In C. Unseld, D. Morrison, D. Sills, and C. Wolf, eds., *Sociopolitical Efficient Energy Use Policy.* Washington D.C.: National Academy of Sciences.
California Energy Commission
  1986   *Residential Appliance Saturation Survey: Pacific Gas and Electric Company.* Sacramento: California Energy Commission.
Cottrell, F.
  1955   *Energy and Society: The Relation Between Energy, Social Change and Economic Development.* New York: McGraw-Hill.
Cowan, R.S.
  1984   *More Work For Mother: The Ironies of Household Technology from the Open Hearth to the Microwave.* New York: Basic Books.
  1989   The consumption junction: A proposal for research strategies in the sociology of technology. In W. Bijker, T.P. Hughes, and T. Pinch, eds., *The Social Construction of Technological Systems.* Cambridge, Mass.: M.I.T. Press.
Cramer, J.C., N. Miller, P. Craig, B. Hackett, T.M. Dietz, M. Leoine, and E. Vine
  1985   Social and engineering determinants and their equity implications in residential electricity use. *Energy* 10(12):1283-91.
Dillman, D.A., E.A. Rosa, and J.J. Dillman

1983   Lifestyle and home energy conservation in the U.S. *Journal of Economic Psychology*
         3:299-315.
Douglas, M., and B. Isherwood
1979   *The World of Goods.* New York: W.W. Norton.
Duncan, O.D.
1978   Sociologists should reconsider nuclear energy. *Social Forces* 57:1-22.
Durning, A.
1992   *How Much is Enough? The Consumer Society and the Future of the Earth.* New York:
         W.W. Norton.
Ewen, S.
1976   *Captains of Consciousness: Advertising and the Social Roots of the Consumer Culture.*
         New York: McGraw-Hill.
1988   *All Consuming Images: The Politics of Style in Contemporary Culture.* New York:
         Basic Books.
Featherstone, M., ed.
1990   *Global Culture: Nationalism, Globalization and Modernity.* London: Sage.
1991   *Consumer Culture and Postmodernism.* London: Sage.
Forty, A.
1986   *Objects of Desire: Design and Society 1750-1980.* London: Thames and Hudson.
Frey, C.J., and D.G. LaBay
1983   A comparative study of energy consumption and conservation across the life
         cycle. In R. Bagozzi and A. Tybout, eds., *Advances in Consumer Research* 10:641-
         646.
Fussell, P.
1983   *Class.* New York: Ballantine Books.
Gartman, D.
1991   Culture as class symbolization or mass reification? A critique of Bourdieu's dis-
         tinction. *American Journal of Sociology* 97:421-447.
Hackett, B., and L. Lutzenhiser
1991   Social structures and economic conduct: Interpreting variations in household en-
         ergy consumption. *Sociological Forum* 6:449-470.
Heberlein, T.A., and G.K. Warriner
1982   The influence of price and attitude on shifting residential electricity consumption
         from on to off-peak periods. *Journal of Economic Psychology* 4:107-130.
Humphrey, C.R., and F.H. Buttel
1982   *Environment, Energy and Society.* Belmont, Calif.: Wadsworth.
Kohno, R.
1984   Energy attitudes and behavior: A cultural comparison of Japanese and American
         families in the United States. Pp. 423-436 in B. Morrison and W. Kempton, eds.,
         Families and Energy: Coping with Uncertainty. East Lansing: Michigan State Uni-
         versity.
Lacy, M.
1985   Apparent and genuine affluence: Their relation to energy consumption. *Sociologi-
         cal Perspectives* 28:117-143.
Lutzenhiser, L., and B. Hackett
1993   Social stratification and environmental degradation: Understanding household
         CO_2 production. *Social Problems* 40:50-73.
Lutzenhiser, L.
1992a   A cultural model of household energy consumption. *Energy—The International
         Journal* 17:47-60.

1992b Modeling Energy Consumption: Social Contexts and Social Myths. Paper presented at annual meeting of the American Sociological Association, Pittsburgh, Pa.

1993 Social and behavioral aspects of energy use. *Annual Review of Energy and the Environment* 18:247-289.

1994 Sociology, energy and interdisciplinary environmental science. *The American Sociologist* 25:57-78.

Lynes, R.
1955 *The Taste-Makers.* New York: Harper and Brothers.

McCracken, G.
1988 *Culture and Consumption: New Approaches to the Symbolic Character of Consumer Goods and Activities.* Bloomington: Indiana University Press.

Mason, R.S.
1981 *Conspicuous Consumption: A Study of Exceptional Consumer Behavior.* New York: St. Martin's Press.

Mazur, A., and E.A. Rosa
1974 Energy and lifestyle. *Science* 607-610.

Miller, D.
1987 *Material Culture and Mass Consumption.* Oxford, England: Basil Blackwell.

Mukerji, C.
1983 *From Graven Images: Patterns of Modern Materialism.* New York: Columbia University Press.

Newman, D.K., and D. Day
1975 *The American Energy Consumer.* Cambridge, Mass.: Ballinger.

Olsen, M.E.
1991 The energy consumption turnaround and socioeconomic well-being in industrial societies in the 1980s. In L. Freese, ed., *Advances in Human Ecology.* Greenwich, Conn.: JAI Press.

Otnes, P., ed.
1988 *The Sociology of Consumption: An Anthology.* New Jersey: Humanities Press.

Packard, V.
1959 *The Status Seekers.* New York: D. McKay Co.

Rosa, E.A., G. Machlis, and K. Keating
1988 Energy. *Annual Review of Sociology* 14:149-172.

Saunders, P.
1990 *A Nation of Home Owners.* London: Unwin Hyman.

Schnaiberg, A.
1991 The Political Economy of Consumption: Ecological Policy Limits. Paper presented at annual meeting of American Association for the Advancement of Science, Washington, D.C.

Schwartz, D., and B. True
1990 What households do when electricity prices go up: An econometric analysis with policy implications. Pp. 2.121-2.130 in *Proceedings of the American Council for an Energy-Efficient Economy.* Washington, D.C.: ACEEE Press.

Skumatz, L.A.
1988 Energy-related differences in residential target-group customers: Analysis of energy usage, appliance holdings, housing, and demographic characteristics of residential customers. Pp. 11.131-11.143 in *Proceedings of the American Council for an Energy Efficient Economy.* Washington, D.C.: ACEEE Press.

Stern, P.C., ed.
   1984   *Improving Energy Demand Analysis*. Panel on Energy Demand Analysis, Commit-
          tee on Behavioral and Social Aspects of Energy Consumption and Production,
          National Research Council. Washington. D.C.: National Academy Press.
Stern, P.C.
   1986   Blind spots in policy analysis: What economics doesn't say about energy use.
          Journal of Policy Analysis and Management 5:200-227.
   1993   A second environmental science: Human-environment interactions. *Science*
          260:1897-1899.
Stern, P.C., E. Aronson, J.M. Darley, D.H. Hill, E. Hirst, W. Kempton, and T.J. Wilbanks
   1986   The effectiveness of incentives for residential energy-conservation. *Evaluation Re-
          view* 10:147-76.
Stern, P.C., and E. Aronson, eds.
   1984   *Energy Use: The Human Dimension*. Committee on Behavioral and Social Aspects
          of Energy Consumption and Production, National Research Council. New York:
          Freeman.
Throgmorton, J.A., and M.J. Bernard, III
   1986   Minorities and energy: A review of recent findings and a guide to future re-
          search. Pp. 7.259-7.280 in *Proceedings of the American Council for an Energy Efficient
          Economy*. Washington, D.C.: ACEEE Press.
Uusitalo, L., ed.
   1983   *Consumer Behavior and Environmental Quality: Trends and Prospects*. New York: St.
          Martin's Press.
Veblen, T.
   1979/ *The Theory of the Leisure Class*. New York: Penguin Books.
   1899
Warde, A.
   1990   Introduction to the sociology of consumption. *Sociology* 24:1.
Weber, M.
   1978   *Economy and Society*. C. Wittich and G. Roth, eds. Berkeley, Calif.: University of
          California Press.
White, L.
   1975   *The Concept of Cultural Systems: A Key to Understanding Tribes and Nations*. New
          York: Columbia University Press.

ENVIRONMENTAL IMPACTS OF POPULATION AND CONSUMPTION

Thomas Dietz and Eugene A. Rosa

How can we investigate the driving forces of environmentally significant consumption? The brief essays in this chapter offer a variety of theoretical and methodological answers to this question. As with most problems in science, multiple strategies will yield a more robust understanding than any single approach. In this essay, we propose the use of statistical models of the driving forces of environmental change to understand impact over time and across nations. This approach is based on a venerable and robust tradition in the social sciences—that of macro-comparative research (Bollen et al., 1993).

To link this social science tradition to work in the environmental sciences, we begin with the IPAT (Impact = Population × Affluence × Technology) equation. The IPAT framework is useful for thinking about the human actions, including consumption, that drive environmental change. The formulation was first offered by Ehrlich and Holdren (Ehrlich and Holdren, 1971, 1972; Holdren and Ehrlich, 1974) in a debate with Commoner (1972a, 1972b) and has seen broad usage since then (Dietz and Rosa, 1994). IPAT is easily understood, frequently used for illustrative purposes and can discipline our thinking. It also serves as a good starting point for a statistical model of the driving forces of environmental change (Dietz and Rosa, 1994, 1997; Preston, 1995). The classical IPAT formulation is an accounting identity and thus must assume rather than test the effects of driving forces on environmental change. In contrast, the statistical model can be used to test hypotheses about driving forces.

The IPAT formulation is simply:

$$I = P \times A \times T \tag{1}$$

where I is environmental impact, P is population, and A is affluence. The typical measure of A is per capita economic activity, so PA becomes aggregate, or total, economic activity. Measures for I, P, and A are used to calculate T, which is by definition $I/(PA)$ or environmental impact per unit economic activity. Thus T represents not only technology *per se*, but also culture, social organization, and all facets of human life other than population and economic activity.

In previous papers (Dietz and Rosa, 1994, 1997), we suggest that the IPAT idea can be reformulated into a stochastic model:

$$I = aP^b A^c T^d e, \tag{2}$$

where a, b, c and d can be either parameters or more complex functions. In either case, they can be estimated using standard statistical procedures. The key change from the traditional *IPAT* approach is that an independent measure of T must be used—the researcher must specify what is meant by technology rather than solving for T as I/PA. I,P,A, and T can represent either single measured variables or vectors of measured variables. The residual term e represents all variables not explicitly included in the model. This makes the residual both interesting and interpretable. It is the multiplier that represents all effects other than those specified in the model. The model is simple, systematic, and robust: simple because it incorporates key anthropogenic driving forces with parsimony, systematic because it specifies the mathematical relationship between the driving forces and their impacts, and robust because it is applicable to a wide variety of impacts.

One can easily think of more complex formulations. For example, a long tradition of economic and environmental models attempts to explore the complex feedbacks among variables considered in a model. Following that tradition, a reasonable first step in elaborating our model would be to draw on the vast literature on population and development to specify equations linking population and economic growth. The realism of such models can be heightened by adding further equations to describe other causal feedbacks. Models with many equations, parameters and variables are commonplace in the econometric literature. But more elaborate models quickly become opaque and strain the limits of available data. Thus at this early stage in modeling the driving forces of environmental change we believe that the best strategy is one that begins with relatively simple models that are easy to understand and that can be disciplined by existing data sources.

We also note that our formulation represents an advance over the existing literature examining the relationship between affluence and environmental impacts. Most of that literature uses only one independent variable, affluence (Grossman and Krueger, 1995; Selden and Song, 1994; Shafik, 1994). In those models, environmental impact is considered a nonlinear function of affluence, as is the case in our model. But the impact of population on the environment usually is assumed to be directly proportional to population size, as in the *IPAT* accounting equation. Any nonproportional effects of population (increasing or decreasing returns to scale) are ignored. Since most debates about the driving forces of environmental change have focused on the impacts of population, we believe models that do not explore population effects will prove of limited value. Thus our model, despite its admitted limitations, represents a step forward. Our goal here is to show how this model can easily be elaborated to answer interesting questions about consumption and the environment.[1]

APPLYING THE MODEL OF CONSUMPTION

Different ideas about consumption can be accommodated by changing the operational definitions of various terms in the model. A common, if controversial, argument about consumption suggests that humans engage in too much economic activity—use up too much ecological "space" (Daly and Cobb, 1989) or have too large an ecological "footprint" (Wackernagel and Rees, 1996). This definition seems to match what Stern (Chapter 2) proposes as the "ecological" or "physics" definition of consumption. Measures of per capita economic activity, such as per capita gross domestic product, used as A in the model, would be an appropriate specification for this interpretation of consumption. Then PA becomes aggregate economic activity—consumption defined in terms of ecological space.[2]

Consumption can also be taken to mean affluence *per se*. As societies become more affluent they not only consume more, but, in addition, their patterns of consumption may shift. This corresponds roughly to the "sociological" definition of consumption. As noted above, a number of recent papers examine the link between environmental impacts and affluence measured as gdp, that is, gross domestic product per capita (Grossman and Krueger, 1995; Holtz-Eakin and Selden, 1995; Selden and Song, 1994; Shafik, 1994). They find a "Kuznets" curve in which increasing affluence (consumption in this definition) leads to increasing impact until a turnover point is reached (usually around $5000-$10,000 in per capita gross domestic product) at which point impact levels out or declines.[3]

Economics offers a more restricted definition of household consumption, defining it as the total spending on consumer goods (Samuelson and Nordhaus, 1989:969). This definition closely matches popular concerns with consumerism and implies that as consumer spending goes up, so

---

[1]Dietz and Rosa (1994) offer a more extended discussion of modeling strategies in the study of environmental impacts.

[2]The models we are specifying here do not take account of the spatial intensity of human activity. But when spatial intensity is an important consideration, as it is when land use is changed, it is simple to add a term to the model representing land area. For an example that applies our formulation to the problem of tropical deforestation, see Dietz et al. (1991).

[3]Unfortunately, as noted above, these analyses all presume directly proportional effects of population (in terms of our model they assume that b = 1). As a result, they may misestimate the effects of both population and affluence.

does environmental impact. In our formulation, consumption defined as consumer spending would be modeled as a term that represents the percent of gross domestic product spent on consumer goods (C):

$$I = aP^bA^cC^fT^de \qquad (3)$$

Then the function represented by f would indicate the importance of consumer spending in generating environmental impact while holding overall affluence constant. We also note, following Cramer (1995), that households rather than individuals are often the key consumption units. This distinction would suggest breaking P into two terms, one for the number of households and one for average household size:

$$I = aS^gH^hA^cC^fT^de \qquad (4)$$

where H is the number of households and S is the average household size.

An example may illustrate the utility of the model. We have examined CO_2 emissions (in millions of metric tons of carbon per year) as a measure of I. The analysis is based on 1989 data for 111 nations. We use population size for P and gross domestic product per capita (gdp) for A. In this formulation, T is combined with e, a unique term for each nation in the analysis that combines the effects of culture, institutions, and technology *per se*. That is, e represents "everything else." Thus the equation estimated is the following:

$$I = aP^bA^ce \qquad (5)$$

It disaggregates consumption into three components: population size (the number of people consuming), affluence (per capita consumption in the sociological sense), and everything else. Details of our analysis are described in Dietz and Rosa (1997).

The effects of population are displayed in Figure 4-2. Population has a strong impact, and there is some evidence of diseconomies of scale in that there are disproportionately large effects for the most populous nations. These results embarrass the argument that population has little effect, or even a beneficial effect, on the environment and lend support to ongoing concern with population growth as a driving force of environmental impacts. Of course, these conclusions are conditional on the cases used in the analysis.

Figure 4-3 indicates that the effects of affluence on CO_2 emissions level off and even decline somewhat at the very highest levels of gross domestic product per capita (gdp). We suspect that this shift is the result of structural changes in both consumption and production, including a

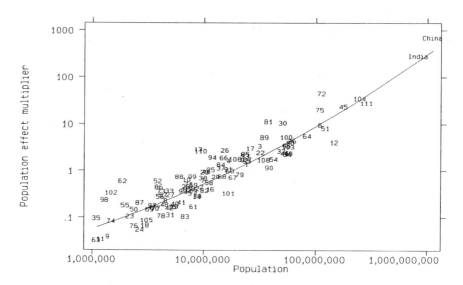

FIGURE 4-2 Effects of population on CO_2 emissions. Solid line represents the effect of population size relative to the geometric mean (12.3 million). Population effects are calculated at the geometric mean of gross domestic product ($1476). The curve reflects the best-fitting log-polynomial model, which has a linear regression coefficient of 1.123 (SE = 0.058), and a quadratic coefficient of 0.063 (SE = 0.026). The numbers in the body of the figure represent the countries used in the analysis; a key is provided in Dietz and Rosa (1997). SOURCE: Dietz and Rosa (1997).

shift to a service-based economy and the ability of the more affluent economies to invest in energy efficiency. (These are hypotheses that can be tested by adding the appropriate indicators to the model.) This decline in impact only occurs when per capita affluence is above $10,000.[4] Seventy-five percent of the 111 nations in our sample have gdps below $5000. Thus our results suggest that for the overwhelming majority of nations, economic growth that can be anticipated for the next quarter century or so will produce production and consumption patterns that lead to increasing rather than declining CO_2 emissions per unit GDP. Reductions in CO_2

---

[4]As noted above, most economic analyses of economic growth and environment, which do not allow for nonproportional effects of population, suggest that impact declines somewhere between $1,000 and $10,000 in per capita GDP (Grossman and Krueger, 1995; Shafik, 1994). The exception is Holtz-Eakin and Selden (1995) whose analysis of CO_2 emissions implies a turning point of over $35,000 per capita.

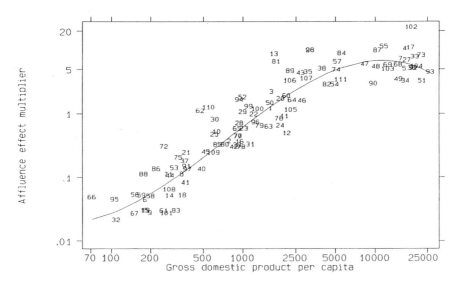

FIGURE 4-3 Effects of affluence on CO_2 emissions. Solid line represents the effect of population size relative to the geometric mean ($1476). Affluence effects are calculated at the geometric mean of population (12.3 million). The curve reflects the best-fitting log-polynomial model, which has a linear regression coefficient of 1.484 (SE = 0.105), a quadratic coefficient of –0.152 (SE = 0.026), and a cubic coefficient of –0.070 (SE = 0.020). The numbers in the body of the figure represent the countries used in the analysis; a key is provided in Dietz and Rosa (1997). SOURCE: Dietz and Rosa (1997).

emissions will not occur in the normal course of development and will have to come from targeted efforts to shift toward less carbon-intensive technologies and activities.

One advantage of our approach is that it is easy to use the model to make projections under alternative scenarios. This allows comparisons with more complex models and can be used to assess policy options. Here, as an example, we use the estimated coefficients of our model to project global CO_2 emissions for the year 2025. In one scenario, we assume that national technological multipliers (the e term) will not change over time. In the second scenario, we assume an increase in efficiency, and thus a decrease in the technology multiplier of 1 percent per year. In this assumption consumption, defined as affluence, increases but changes in production processes and the bundle of goods and services consumed lead to less impact per unit consumption. In both cases, we use the United Nations medium-case scenario for population projections and assume a 2 percent annual real growth in gdp (World Resources Institute, 1992). The first scenario, with no technological progress, implies global CO_2 emissions in 2025 of 4.3×10^{10} metric tons, a 95 percent increase over 1991 emissions. A 1 percent per year increase in carbon efficiency would mean an increase of only 36 percent, to 3.0×10^{10} metric tons. To achieve a goal of stable emissions at 1991 levels in the face of growth in consumption and population, our model suggests efficiency increases would need to average about 1.8 percent per year from 1990 to 2025. While such increases are feasible, they will not occur without strenuous efforts.

FURTHER DIRECTIONS

In the interest of clarity, we have avoided complexity in this analysis. Of course, considering only two candidate driving forces, population and affluence, is not adequate for understanding environmental change. As noted above, our model can easily be expanded to assess ever more subtle and detailed hypotheses about the effects of consumption and other driving forces on the environment. Equation 4 above, for example, allows an examination of the relative contribution of number of households, average household size (a disaggregation of population into two components), gross domestic production (gdp) per capita, and percent of GDP spent on consumption (a disaggregation of overall GDP into consumption as defined by economists). As we have noted elsewhere (Dietz and Rosa, 1994), the method allows work on driving forces of global change to link directly with the substantial body of methodological, theoretical, and empirical work on macro-comparative data in the social sciences. In doing so, it provides a macro complement to analyses such as those of Schipper (Chapter 3) that disaggregate by usage or those of Duchin (Chapter 3) or Lutzenhiser (Chapter 4) that focus on the micro-level of households and individuals.

ACKNOWLEDGMENT

We thank W. Catton, R. Dunlap, A. Ford, E. Franz, L. Hamilton, L. Kalof, and P. Stern for their comments. This work was supported in part by National Science Foundation Grants SES-9109928 and SES-9311593, by the Dean of the College of Liberal Arts at Washington State University, and by the International Institute of George Mason University. Figures 4-2 and 4-3 were originally published in Dietz and Rosa (1997).

REFERENCES

Bollen, K.A., B. Entwistle, and A.S. Alderson
    1993   Macrocomparative research methods. *Annual Review of Sociology* 19: 321-351.

Commoner, B.
  1972a A bulletin dialogue on "The Closing Circle": Response. *Bulletin of the Atomic Scientists* 28(5):17, 42-56.
  1972b The environmental cost of economic growth. Pp. 339-363 in R.G. Ridker, ed., *Population, Resources and the Environment*. Washington, D.C.: Government Printing Office.
Cramer, J.C.
  1995 Population Growth and Air Quality in California. Paper presented at the Eighth Conference of the Society for Human Ecology, South Lake Tahoe, Calif., October 13.
Daly, H.E., and J.B. Cobb, Jr.
  1989 *For the Common Good: Redirecting the Economy Toward Community, the Environment and a Sustainable Future*. Boston: Beacon Press.
Dietz, T., L. Kalof, and R.S. Frey
  1991 On the utility of robust and resampling estimators. *Rural Sociology* 56:461-474.
Dietz, T., and E.A. Rosa
  1994 Rethinking the environmental impacts of population, affluence and technology. *Human Ecology Review* 1:277-300.
  1997 Effects of population and affluence on CO_2 emissions. *Proceedings of the National Academy of Sciences* 94(1):175-179.
Ehrlich, P.R., and J.P. Holdren
  1971 Impact of population growth. *Science* 171:1212-1217.
  1972 Impact of population growth. Pp. 365-377 in R.G. Ridker, ed,. *Population, Resources and the Environment*. Washington, D.C.: U.S. Government Printing Office.
Grossman, G., and A. Krueger
  1995 Economic growth and the environment. *Quarterly Journal of Economics* :353-377.
Holdren, John P., and Paul R. Ehrlich
  1974 Human population and the global environment. *American Scientist* 62:282-292.
Holtz-Eakin, D., and T.M. Selden
  1995 Stoking the fires? CO_2 emissions and economic growth. *Journal of Public Economics* 57:85-101.
Preston, S.H.
  1995 The effect of population growth on environmental quality. *Population Research and Policy Review*.
Samuelson, P.A., and W.D. Nordhaus
  1989 *Economics*. New York: McGraw-Hill Book Company.
Selden, T.M., and D. Song.
  1994 Environmental quality and development: Is there a Kuznets curve for air pollution emissions? *Journal of Environmental Economics and Management* 27:147-162.
Shafik, N.
  1994 Economic development and environmental quality: An econometric analysis. *Oxford Economic Papers* 46:757-773.
Wackernagel, M., and W. Rees
  1996 *Our Ecological Footprint*. Gabriola Island, B.C., Canada: New Society Publishers.
World Resources Institute
  1992 *World Resource, 1992-1993*. New York: Oxford University Press.

CROSS-NATIONAL TRENDS IN FOSSIL FUEL CONSUMPTION, SOCIETAL WELL-BEING, AND CARBON RELEASES

Eugene A. Rosa

RESEARCH QUESTIONS

This volume (see Chapter 2) provides a comprehensive, environmentally sensitive definition of consumption as consisting ". . . of human and human-induced transformations of materials and energy. Consumption is environmentally important to the extent that it makes materials or *energy less available* for future use, *moves a biophysical system* toward a different state or, through its effects on those systems, threatens human *health, welfare, or other things people value"* (emphases added). The research reported here addresses two classes of threats contained in this definition: threats to atmospheric systems and threats to human well-being.

Fossil fuel consumption has been the foundation of industrial production and modernity for well over a century. This consumption is pivotal in environmental importance because continued consumption of its finite stocks will make it less available for future use. But it is also pivotal because it is the principal anthropogenic source of the trace greenhouse gas, carbon dioxide (CO_2). Carbon emissions are the smoking gun of fossil fuel consumption. Continued growth in the consumption of carbon-based fuels threatens to move the atmospheric system toward a warmer state. What have been the recent historical patterns of fossil fuel use and of the resulting CO_2[1] loads for the leading industrial nations—the dominant consumers of fossil fuels and producers of CO_2?

Climate change due to continuously increasing CO_2 loads could, in turn, threaten human health, well-being, or other features of social life. How can we measure well-being in a way that neither ignores its multiple domains nor relies on a single domain and single indicator, such as is the practice with aggregate economic measures? How can we assess whether the threats to well-being of increased CO_2 are being realized?

---

[1]CO_2 releases are due primarily to fossil fuel combustion and secondarily to the loss of moist forests. In the case of the leading industrial nations, releases are due almost entirely to fossil fuel use. National estimates of CO_2 emissions for the leading industrial nations, such as those presented here, are computed by the Carbon Dioxide Information and Analysis Center (CDIAC) at Oak Ridge National Laboratory (Boden et al., 1990) by applying a carbon conversion formula to national levels of fossil fuel consumption taken from United Nations compilations and then adding the minuscule amounts of CO_2 due to cement production and gas flaring.

Taken together the three foregoing questions converge to the central issue addressed here: to what extent is there an historical relationship between the two concomitants of industrialization, CO_2 loads and well-being? More specifically, is there a coupling between fossil fuel consumption (or CO_2 emissions) and well-being? Data from the 1970-1985 period clearly show that energy and economic activity, one domain of well-being, had decoupled in several wealthy industrialized countries. But what about other domains of well-being?[2] Along with the carbon emissions it produces, industrialization has provided not only clear economic advantages to societies but also social and other advantages as well. The enjoyable features of modern lifestyles—characterized by the availability of a broad array of goods and services, by a remarkable geographical mobility, by the power of personal climate control, as well as other things people value—depend largely upon industrial production and upon energy. Are these benefits coupled with fossil fuel use via industrialization?

Knowing whether noneconomic well-being is tightly or loosely coupled to fossil fuel consumption and CO_2 emissions can inform debates over broad carbon policy. If, unlike the economic domain, these other domains are tightly coupled, then policies calling for a reduction in CO_2 emissions via reduced fossil fuel use will need to anticipate the costs to well-being of pursuing such a policy. On the other hand, if these domains, too, are decoupled from CO_2, then policies to reduce fossil fuel consumption should anticipate no dramatic downturns in well-being. We address this question by examining 35-year trends in CO_2 emissions and societal well-being.

MEASURES OF WELFARE AND WELL-BEING

Well-being is a principal inquiry of the economic sciences. The typical way—indeed, virtually the only way—that aggregate welfare or well-being is measured is with summary measures of the economy derived from national accounts. Thus, measures such as gross national product (GNP) or gross domestic product (GDP) are used to assess welfare.[3] Since

---

[2]These domains, such as health and lifestyle, do not always correlate highly with economic measures (see, for example, Sen, 1993).

[3]Economists have assessed CO_2 impact costs using aggregate economic indicators to address two questions: (i) What is the relationship between national levels of economic activity and CO_2 loads, and (ii) What would be the impacts to economic activity from alternative carbon policies? (See, for example, Nordhaus, 1991; Dowlatabadi and Morgan, 1993; Schelling, 1992; Peck and Teisberg, 1993; and Rothen, 1995, who provides a summary of European efforts along these same lines.)

energy growth partially decoupled from economic growth in the 1970s and 1980s (see, for example, Alterman, 1985), it follows that energy also decoupled from well-being—at least as measured in terms of economic growth.

There are significant, well-known limitations to using a single national product indicator as a measure of welfare. Two inadequacies have been singled out for particular criticism: (i) that GNP and GDP summarize market transactions, not welfare (see for example, Sen, 1982, 1993); and (ii) that these measures ignore the external costs to "natural capital" or "ecosystem services" (Coase, 1960; Nordhaus and Tobin, 1973; Daly and Cobb, 1989; Daly and Townsend, 1992; Costanza, 1995, and a variety of important critical chapters in Costanza, 1991).

With the foregoing considerations in mind our objective was to develop a more comprehensive measure of well-being, one that captured important domains of social life left unaddressed with economic measures. To meet this objective we first identified the world's 23 leading industrial nations.[4] We then subjected 26 readily available cross-national indicators representative of key domains of social life—health, nutrition, transportation, education, culture, communications and media, and general satisfaction—to a principal factor analysis.[5] The result was a four-factor solution, comprising 23 of the original 26 indicators, and representing four broad domains of social life: modern lifestyle, general well-being, health and safety, and life stress. Results of the factor analyses are presented in Table 4-5.

MONITORING TRENDS IN CO_2 LOADS AND SOCIETAL WELL-BEING

To determine the degree of parallelism, or coupling, between CO_2

---

[4]We confined the analyses to the leading industrial nations because they produce the lion's share of world CO_2 loads, they have the greatest policy flexibility, and it is there that we find generally reliable data of the type needed for this analysis.

[5]An alternative approach would be to rely on "subjective" measures of social well-being, thereby tapping into peoples' perceptions of their life experiences. We did not pursue this option on practical and substantive grounds. First, unlike social indicator data that are routinely collected by international agencies, subjective "quality-of-life" data are only collected episodically and not always for a consistent set of countries. Second, an earlier literature was consistent in showing little relationship between the "objective" conditions of life and subjective satisfaction (Campbell, 1981; Campbell et al., 1976). A more recent literature, though questioning this long-standing conclusion (see, for example, Veenhoven, 1991, 1988) is inconclusive about the relationship between objective and subjective indicators of well-being, an ambiguity even more pronounced when focusing exclusively on the leading industrial nations (Myers and Diener, 1995), as is the case here.

TABLE 4-5 Four-Factor Model: Rotated Solution

	Factor Loading	Eigenvalue	Proportion of Variance Explained
Modern Life Style		10.24	.58
Televisions	.86		
Divorce rate	.83		
Radio receivers	.82		
University students	.79		
Cars	.76		
Telephones	.75		
Commercial vehicles	.71		
Secondary students	.58		
Cinemas	−.40		
Hours worked per week	−.46		
General Well-Being		2.23	.13
Life expectancy—males	.81		
Life expectancy—females	.76		
Books published	.67		
Daily protein supply	.57		
Physicians	.47		
Infant mortality rate	−.69		
Health and Safety		1.75	.10
Daily food supply	−.83		
Daily fat supply	−.81		
Cancer deaths	−.75		
Diabetes deaths	−.68		
Accident deaths (autos)	−.62		
Life Stress		1.10	.06
Pharmacists	−.59		
Ulcer deaths	.51		
Total Model		.86	

loads (driven by fossil fuel consumption) and societal well-being, we examined trends in CO_2 loads and the social well-being measures for the period 1950-1985 for the 23 industrial nations in our data set. We first examined the trend in per capita yearly CO_2 emissions for each of the 23 countries, expecting to find them all on a consistently increasing, monotonic trend until such time as economic decoupling took place. This was the case until 1970. We then found (verified by a formal nonparametric test using the Mann-Kendall statistic) the evolution of 3 distinct trends after 1970: an increasing trend for 6 nations (Australia, Greece, Italy, Norway, Portugal, and Spain), a stabilizing trend for 10 nations (Austria,

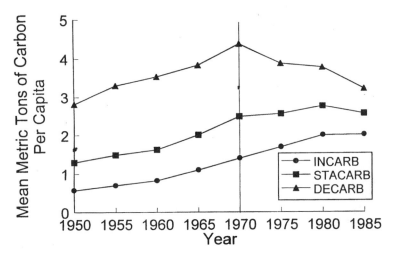

FIGURE 4-4 Carbon emissions per capita: All groups. INCARB = increasing trend in per capita CO_2 emissions; STACARB = stabilizing trend in per capita CO_2 emissions; DECARB = decreasing trend in CO_2 emissions.

Canada, Denmark, Finland, Germany, Iceland, Ireland, Japan, the Netherlands, and New Zealand), and a decreasing trend for seven nations (Belgium, France, Luxembourg, Sweden, Switzerland, the United Kingdom, and the United States).

We label the three trends INCARB, STACARB, and DECARB, respectively. Plots of the mean carbon loads for the groups of countries within each trend are presented in Figure 4-4.

To assess whether carbon loads parallel the social aspects of well-being, we plotted summated scores based upon our factor analysis for each of the four domains of the factor solution: modern lifestyle, general well-being, health and safety, and life stress (Figure 4-5). The summated scores, by carbon grouping, are all either monotonically increasing or monotonically decreasing. With some exception for the economic laggards, the INCARB countries, we find that the social domains of well-being do not closely parallel the three distinct carbon trends. It appears that the coupling of well-being, whether defined narrowly or broadly, to fossil fuel consumption and the carbon emissions it produces weakened considerably in the 1970-1985 period.

We interpret these findings with a Threshold-Asymptote-Decoupling (TAD) hypothesis. A threshold level of energy consumption is a prerequisite for a nation to reach industrial status, but beyond that threshold there is considerable flexibility in the amounts of energy needed to sustain or improve standards of well-being. Noneconomic measures of well-

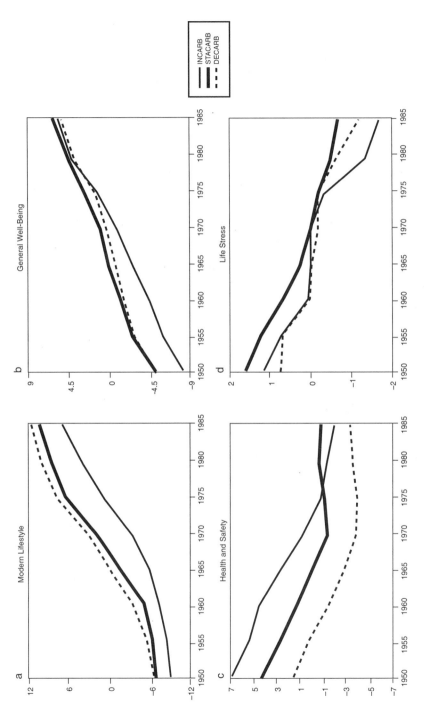

FIGURE 4-5 Trends in four dimensions of societal well-being: Summated scores, all groups of countries, 1950-1985. INCARB = increasing trend in per capita CO_2 emissions; STACARB = stabilizing trend in per capita CO_2 emissions; DECARB = decreasing trend in CO_2 emissions.

being of the type examined here, such as life expectancy, food intake, and types of death, have inherent limitations, and therefore are expected to reach some asymptote. Thus, the decoupling of energy from well-being is, in part, due to the asymptotic limitations in the well-being measures.

Figure 4-6 presents the relationships of carbon emissions to well-being in more detail and suggests a complex picture. Generally, the correlations are of much larger magnitude in the INCARB group than in the other groups, particularly in the earlier years of the time series. This finding is consistent with the TAD hypothesis. However, the correlations do not diminish with time for all groups and all indicators of well-being or drop sharply after 1970 in the DECARB countries, as might be expected if well-being was decoupling from carbon emissions during this period. Rather, the trends seem to move in different directions across indicators and country groups. These results may be interpretable in terms of properties of the particular indicators or of particular countries within groups (because the groups are small, a single anomaly can have a large effect on correlations). The proper interpretation awaits a finer-grained analysis.

SUMMARY OF RESULTS

We have used historical trends to ask whether there is a parallelism between carbon loads and societal well-being and how closely key indicators of well-being couple with carbon loads. We found, in general, that well-being after 1970 did not closely track trends in per capita carbon loads in 17 of the 23 societies examined (the DECARB and STACARB groups). Parallelisms did remain between carbon loads and our well-being indicators for the INCARB group of six countries—a positive parallelism for the economic measure[6] and the modern lifestyle and general well-being measures, on the one hand, but an inverse parallelism for our health and safety and stress measures, on the other.

Because the indicators examined are well-behaved over a sizable spell of recent history (35 years) and because the types of indicators examined are not prone to abrupt or precipitous change, we can derive reasonable substantive and policy conjectures from our results. The patterns we are observing, we hypothesize, reflect a structural transformation of the advanced industrial societies: from an industrial-based modernity to a service- and communication-based postmodernity.[7] The three separate pat-

---

[6]These data are included as part of our larger analyses but are not included here for lack of space.

[7]A number of European scholars (see, for example, Mol, 1995) are promoting a theory of "ecological modernization"—whose main features are consistent with our results—to describe the next stage of development for industrial societies.

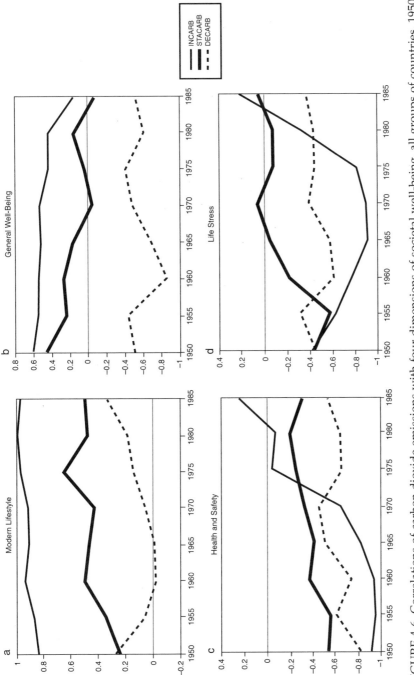

FIGURE 4-6 Correlations of carbon dioxide emissions with four dimensions of societal well-being, all groups of countries, 1950-1985. INCARB = increasing trend in per capita CO_2 emissions; STACARB = stabilizing trend in per capita CO_2 emissions; DECARB = decreasing trend in CO_2 emissions.

terns of carbon loads (INCARB, STACARB, DECARB), then, reflect different stages in this transformation. On the policy side, we conjecture that broad strategies devoted to the reduction of CO_2 emissions via reduced fossil fuel use in the leading industrial nations will probably not be accompanied by significant impacts to societal well-being—at least, not over the next several decades.

This conjecture presumes that reductions in fossil fuel consumption can continue to come from technical improvements and other innovations that do not interfere with material well-being, as apparently occurred between 1970 and 1985. There is some evidence, however, that the pattern of energy use that produced decoupling during that period has been changing. For instance, Schipper (Chapter 3) identifies ways that energy use has been shifting "from production to pleasure," a shift that may make it more difficult to achieve future energy and emissions reductions without reducing well-being. Detailed analysis of data on carbon and well-being since 1985 can help address this issue.

STRENGTHS AND SHORTCOMINGS OF THIS APPROACH

The principal strength of the approach outlined here is that it provides a way of addressing an essential analytic and policy question through an aggregate conceptualization of social welfare and with readily available archival data. Furthermore, it represents an effort to broaden the concept of well-being in a way that more thoroughly reflects the threats consumption may pose to human health, welfare, and other things people value. By differentiating the concept of well-being, it raises questions that suggest promising directions for further research. The approach has two principal weaknesses. First, as with all aggregate measures of human activity, the aggregate data mask considerable underlying detail and countervailing processes. Second, our approach is more empirically driven than theory driven with the result that it can identify relationships with considerably greater ease than it can explain them.

REFERENCES

Alterman, J.
    1985   *A Historical Perspective on Changes in the U.S. Energy-Output Ratios.* Washington, D.C.: Resources for the Future.
Boden, T.A., P. Kanciruk, and M.P. Farrell
    1990   *Trends '90: A Compendium of Data on Global Change.* Carbon Dioxide Information Analysis Center. Oak Ridge, Tenn.: Oak Ridge National Laboratory.
Campbell, A.
    1981   *The Sense of Well-Being in America: Recent Patterns and Trends.* New York: McGraw-Hill.

Campbell, A., P. Converse, and W. Rodgers
    1976   *The Quality of American Life: Perceptions, Evaluations, and Satisfactions.* New York: Russell Sage Foundation.
Coase, R.
    1960   The problem of social costs. *Journal of Economics and Law* 3:1-44.
Costanza, R.
    1991   *Ecological Economics: The Science and Management of Sustainability.* New York: Columbia University Press.
    1995   Integrated Ecological Economic Modeling and Adaptive Management of Complex Systems. Lecture, annual meetings of the American Association for the Advancement of Science, Atlanta, Ga. February 17.
Daly, H.E., and J.B. Cobb, Jr.
    1989   *For the Common Good: Redirecting the Economy Toward Community, the Environment, and a Sustainable Future.* Boston: Beacon Press.
Daly, H.E., and K.N. Townsend, eds.
    1992   *Valuing the Earth: Economics, Ecology, Ethics.* Cambridge, Mass.: M.I.T. Press.
Dowlatabadi, H., and M.G. Morgan
    1993   Integrated assessment of climate change. *Science* 259: 1813,1932.
Mol, A.R.J.
    1995   *The Refinement of Production: Ecological Modernization Theory and the Chemical Industry.* The Hague: CIP-Data Konninklijke Bibliotheek.
Myers, D.G., and E. Diener
    1995   Who is happy? *Psychological Science* 6:10-19.
Nordhaus, W.D.
    1991   To slow or not to slow: The economics of the greenhouse effect. *The Economic Journal* 101:920-937.
Nordhaus, W.D., and J. Tobin
    1973   Is growth obsolete? Pp. 509-532 in *The Measurement of Economic and Social Performance.* New York: National Bureau of Economic Research.
Peck, S.C., and T.J. Teisberg
    1993   *Optimal CO_2 Emissions Control with Partial and Full World-Wide Cooperation: An Analysis Using CETA.* Palo Alto, Calif.: Electric Power Research Institute.
Rothen, S.M.
    1995   *The Greenhouse Effect in Economic Modeling: A Critical Survey.* Dübendorf, Switzerland: EWAG (Swiss Federal Institute for Environmental Science and Technology).
Schelling, T.C.
    1992   Some economics of global warming. *American Economic Review* 82:1-14.
Sen, A.
    1982   *Choice, Welfare, and Measurement.* Oxford, England: Basil Blackwell.
    1993   The economics of life and death. *Scientific American* 208(5):40-47.
Stern, P., O.R. Young, and D. Druckman, eds.
    1992   *Global Environmental Change: Understanding the Human Dimension.* Committee on the Human Dimensions of Global Change, National Research Council. Washington, D.C.: National Academy Press.
Veenhoven, R.
    1988   The utility of happiness. *Social Indicators Research* 20:333-354.
    1991   Is happiness relative? *Social Indicators Research* 24:1-34.

EMULATION AND GLOBAL CONSUMERISM

Richard R. Wilk

There are good reasons for concern about the environmental impacts of 5 billion people consuming at the level of the developed countries of Europe and North America. Given high economic growth rates in many parts of the developing world, as well as the rapid spread of electronic media, advertising, and consumer goods, we must ask what kind of consuming future we can expect in areas that are now constrained by poverty and isolation. If everyone develops a desire for the Western high-consumption lifestyle, the relentless growth in consumption, energy use, waste, and emissions may be disastrous.

It is also possible, however, that each country, region, or ethnic group may maintain different aspirations, definitions of living standards, and consumption goals. Then we could expect a high degree of diversity in consumer demand and, perhaps, much more moderate long-term levels of consumption, even with more equal distribution of income. Of course, it is also possible that a large part of the developing world will never achieve the threshold income levels necessary to consume large amounts of durables, luxuries, or services, whatever their aspirations.

The choice of different scenarios for the consumption trajectories of the developing world hinges partially on the issue of emulation. Do people of other cultures find the Western high-consumption model attractive? Or do their own cultures offer strong alternative values that make foreign models less attractive? Are some groups or cultures more likely to emulate the West than others? What cultural, social, and economic forces promote high-consumption lifestyles?

The strongest form of emulation is often labeled "cultural imperialism." This theory contends that a combination of Western control of mass media and improved advertising, along with falling trade barriers and the spread of industrial capitalism, will inevitably lead the developing world into emulative forms of consumption (Tomlinson, 1991; Rassuli and Hollander, 1986). There are various moral positions on cultural imperialism; some see it leading to economic freedom, while others consider it a malign form of brainwashing and false consciousness (Ewen, 1988; Horowitz, 1988). Many social scientists reject cultural imperialism and contend that instead of increasing centralization and homogenization, the next century will be dominated by new forms of nationalism, localism, and cultural fundamentalism that will challenge both the economic and cultural hegemony of the West (Foster, 1991). There is some empirical evidence for both processes; some forms of localization are concurrent with other kinds of globalization; heterogeneity and homogeneity both seem to be increasing in different sectors and at different scales (Friedman, 1990; Hannerz, 1987; Featherstone, 1990; Wilk, 1995; Drummond and Patterson, 1988; Belk and Dholakia, 1995).

Recent historical and social-scientific research on consumption has produced a great deal of empirical data and many excellent case studies that bear directly on the issue of consumer emulation. Studies of the growth of consumer culture in post-World War II Japan (Tobin, 1992), France (Kuisel, 1993), and Austria (Wagnleitner, 1994) all argue that emulation is not at all mechanical or inevitable but is, instead, a temporary product of specific political policies, trade practices, and cultural influences.

The explosion of consumer demand in China during the last decade has been used both as an example of cultural imperialism and of autonomy and increasing diversity in consumer culture. Consumer aspirations have changed several times during the last 20 years, and comparisons show that the mixture of goods consumed in China at a particular level of income is quite different from that found in other Asian countries (Sklair, 1994).

The case studies demonstrate that people, in general, tend to emulate local elites rather than following a single global generic "Western" consumer model. Sometimes people explicitly reject foreign models instead of emulating them. Periods of emulation may alternate with intervals where global or international standards are rejected in favor of local goods or styles (e.g., Andrae and Beckman, 1985; Appadurai, 1986, 1990; Friedman, 1994)

The influence of Western media and the advertising of global brands on actual consumer behavior and aspirations is still not clearly understood, and the "cultivation effect" of television is weak or highly variable in cross-national data (Ware and Dupagne, 1994; Liebes and Katz, 1990; Moore, 1993). The few comparative cross-cultural studies of consumer aspirations and values are difficult to interpret. Technocratic, urban, highly educated groups in many parts of the world show some increasing commonalities in aspirations and cultural beliefs. But are these sectors the leaders of a new wave of consumerism or small Western cultural enclaves of technocratic and academic "cosmopolitans" (Hannerz, 1990; Belk, 1988; James, 1993)? Methodological problems (translation, sampling) also make these survey results equivocal (Holt, 1994). The precise linkage between Westernized values and the consumption of Western goods is also not well established. Economists find that the determinants of national demand, and the amount saved and invested instead of spent on consumer goods, are complex. The share of increased income spent on major consumption categories, such as food, durables, and housing, varies widely at the same income level, as does the savings rate (Lluch et al., 1977; Gereffi and Korzeniewicz, 1994).

Even the historical development and present fluorescence of mass-consumer society in the West is poorly understood, despite a wealth of new studies (e.g., Mintz, 1985; Brewer and Porter, 1993; Cross, 1993; Richards, 1991; Benson, 1994; Tierstin, 1993; Craik, 1994; McCracken, 1988; Csikszentimihaly and Rochberg-Halton, 1981). Historians tend to agree that the growth in Western consumer demand resulted from a breakdown of rigid class hierarchies and a relaxation of religious inhibitions on conspicuous consumption. Some authors stress the emergence of a "romantic ethic" (Campbell, 1987) or a trend toward cultivating health and self-improvement (Lears, 1989). Recent work has shifted away from "social emulation" or class-competition models of consumer demand, toward a focus on communication, nationalism, advertising, and the growth of markets and retailing. Because all of these trends are occurring in developing countries, we can expect a similar growth of consumer culture, even in the absence of any specific form of emulation. So, even without emulation, consumption levels in developing countries may dramatically increase.

Judging the emulation hypothesis is premature because social science, in general, still lacks a well-established general theory of consumption. Each discipline tends to focus on consumption from its own narrow perspective (though see Miller, 1995). More synthetic work is being done in hybrid fields like consumer research, cultural studies, and social history (Sherry, 1995). Studies of household decision making in several disciplines have been especially promising, because most major investment and consumer decisions are made at the household level. There is considerable constructive debate on models of intra-household bargaining, gender, and power that have direct implications for understanding consumption, spending, and savings behavior (e.g., Phipps and Burton, 1995; Folbre, 1994).

Given the importance of predicting future global demand for consumer goods, energy, water, food, and other resources, we need to better establish the determinants of consumer behavior. While there is a good deal of empirical data available that bear on these issues, there has been little cross-disciplinary synthesis, and fundamental theoretical issues remain unresolved. The major points that emerge from the literature in several disciplines include the following:

(1) There is still no generally accepted model of consumer behavior.
(2) The database for cross-cultural comparison of consumption is poor in quality.

(3) No single academic discipline has adequate tools or data for studying cross-cultural consumer behavior.

(4) The development of consumer culture in developing countries is following a different trajectory from the historical path of the West.

(5) There is still every reason to think that consumption will increase as incomes rise, but we cannot yet predict how that increase will be apportioned to various goods or sectors.

(6) Simple emulation remains an empirically weak model for prediction.

REFERENCES

Andrae, G., and B. Beckman
  1985   *The Wheat Trap: Bread and Underdevelopment in Nigeria*. London: Zed Books.
Appadurai, A., ed.
  1986   *The Social Life of Things*. Cambridge, England: Cambridge University Press.
  1990   Disjuncture and difference in the global cultural economy. *Theory, Culture, and Society* 7:295-310.
Belk, R.
  1988   Third world consumer culture. *Research in Marketing*. Supplement 4.:103-127.
Belk, R., and N. Dholakia
  1995   *Consumption and Marketing: Macro Dimensions*. Belmont, Calif.: Southwestern.
Benson, J.
  1994   *The Rise of Consumer Society in Britain, 1880-1980*. London: Longman.
Brewer, J., and R. Porter, eds.
  1993   *Consumption and the World of Goods*. New York: Routledge.
Campbell, C.
  1987   *The Romantic Ethic and the Spirit of Modern Consumerism*. Oxford, England: Basil Blackwell.
Craik, J.
  1994   *The Face of Fashion*. London: Routledge.
Cross, G.
  1993   *Time and Money: The Making of Consumer Culture*. London: Routledge.
Csikszentimihalyi, M., and E. Rochberg-Halton
  1981   *The Meaning of Things: Domestic Symbols and the Self*. Chicago: University of Chicago Press.
Drummond, P., and R. Patterson, eds.
  1988   *Television and its Audience: International Research Perspectives*. London: British Film Institute Publishing.
Ewen, S.
  1988   *All Consuming Images*. New York: Basic Books.
Featherstone, M., ed.
  1990   *Global Culture: Nationalism, Globalization and Modernity*. London: Sage Publications.
Folbre, N.
  1994   *Who Pays for the Kids? Gender and the Structure of Constraint*. London: Routledge.
Foster, R.
  1991   Making national cultures in the global ecumene. *Annual Review of Anthropology* 20:235-260.

Friedman, J., ed.
  1994   *Consumption and Identity.* Chur, Switzerland: Harwood.
  1990   Being in the world: Globalization and localization. *Theory, Culture and Society* 7:311-328.
Gereffi, G., and M. Korzeniewicz, eds.
  1994   *Commodity Chains and Global Capitalism.* Westport, Conn.: Praeger.
Hannerz, U.
  1987   The world in Creolization. *Africa* 57(4):546-559.
  1990   Cosmopolitans and locals in world culture. *Theory, Culture, and Society* 7:237-251.
Holt, D.
  1994   Consumers' cultural differences as local systems of tastes: A critique of the personality/values approach and an alternative framework. *Asia Pacific Advances in Consumer Research* 1:178-184.
Horowitz, D.
  1988   *The Morality of Spending.* Baltimore, Md.: Johns Hopkins University Press.
James, J.
  1993   *Consumption and Development.* New York: St. Martin's Press.
Kuisel, R.
  1993   *Seducing the French: The Dilemma of Americanization.* Berkeley: University of California Press.
Lears, T.J.J.
  1989   Beyond Veblen: Rethinking consumer culture in America. Pp. 73-97 in Simon Bronner, ed., *Consuming Visions: Accumulation and Display of Goods in America, 1880-1920.* New York: Norton.
Liebes, T., and E. Katz
  1990   *The Export of Meaning: Cross-Cultural Readings of Dallas.* New York: Oxford University Press.
Lluch, C., A. Powell, and R. Williams
  1977   *Patterns in Household Demand and Saving.* New York: Oxford University Press and the World Bank.
McCracken, G.
  1988   *Culture and Consumption.* Bloomington: Indiana University Press.
Miller, D.
  1995   *Acknowledging Consumption.* London: Routledge.
Mintz, S.
  1985   *Sweetness and Power: The Place of Sugar in Modern History.* New York: Penguin Books.
Moore, S.
  1993   *Interpreting Audiences: The Ethnography of Media Consumption.* London: Sage Publications.
Phipps, S., and P. Burton
  1995   Social/institutional variables and behavior within households: An empirical test using the Luxembourg income study. *Feminist Economics* 1(1):151-174.
Rassuli, K. and S. Hollander
  1986   Desire—induced, innate, insatiable? *Journal of Macromarketing* 6(2):4-24.
Richards, T.
  1991   *The Commodity Culture of Victorian Britain: Advertising and Spectacle, 1851-1914.* Stanford, Calif.: Stanford University Press.
Sherry, J., ed.
  1995   *Contemporary Marketing and Consumer Behavior: An Anthropological Sourcebook.* Thousand Oaks, Calif.: Sage Publications.

Sklair, L.
1994   The culture-ideology of consumerism in urban China. *Research in Consumer Behavior*, Vol. 7. Greenwich, Conn.: JAI Press.

Tiersten, L.
1993   Redefining consumer culture: Recent literature on consumption and the bourgeoisie in Western Europe. *Radical History Review* 57:116-159.

Tobin, J., ed.
1992   *Re-Made in Japan: Everyday Life and Consumer Taste in a Changing Society*. New Haven, Conn.: Yale University Press.

Tomlinson, J.
1991   *Cultural Imperialism: A Critical Introduction*. Baltimore, Md.: Johns Hopkins University Press.

Wagnleitner, R.
1994   *Coca-Colonization and the Cold War*. Chapel Hill: University of North Carolina Press.

Ware, W., and M. Dupagne
1994   Effects of U.S. television programs on foreign audiences: A meta-analysis. *Journalism Quarterly* 71(4):947-959.

Wilk, R.
1995   The local and the global in the political economy of beauty: From Miss Belize to Miss World. *Review of International Political Economy* 2(1):117-134.

CULTURAL AND SOCIAL EVOLUTIONARY DETERMINANTS OF CONSUMPTION

Willett Kempton and Christopher Payne

In the workshop paper summarized here (see Kempton and Payne, forthcoming, for the complete version), we draw data from a wide range of human societies to ask: What can cross-cultural comparisons teach us about the relationship between consumption and quality of life? We argue that the dependence of quality of life on consumption is not monotonic and is both weaker and more complex than is often assumed.

We begin by addressing two myths that underlie much thinking about consumption. The first myth is that quality of life generally increases with higher consumption levels—that is, more consumption of goods and services increases the quality of life. This relationship is believed to hold across societies and across social strata within any society. Parts of it are parodied in the tee-shirt slogan "He who dies with the most toys wins." The second myth is that society evolves and changes to improve the lot of individuals. If our society previously had one form of government, kinship system, economy, or whatever and another form replaces it, the societal change improves the quality of life of members of the changed society. We call these the "most toys" myth and the "social evolution for individual benefit" myth. They are addressed at several points here.

We begin by considering the types of societies within which biologically modern humans evolved. These societies are small, organized around family relationships, and subsist by hunting and gathering. Both mobility and their social organization limit consumption. Mobile societies shift residences, whether on a predictable yearly cycle based upon seasonal cycles of wild crops and game or moving more opportunistically to follow herds, water, or areas not yet exhausted of plant resources. Individuals in these societies limit consumption simply via the limit on inventories—you can't possess more than you can carry. Socially, hunting and gathering societies are organized around family relationships and are egalitarian.

We also briefly examine a subsequent form of subsistence, swidden agriculture. This pattern relies on cutting and burning forest, farming the plot for one or a few years, and abandoning it for decades to lie fallow and regrow. As societies moved from hunting and gathering to swidden agriculture, and then to fixed agriculture, changes in social and political organization accompanied these production and settlement changes. Among other things, these changes increase status differentiation. With larger populations in settlements and social differentiation comes the need for display of status by means of prestige goods. Subsequently in social evolution, material consumption is driven partially by status competition. Remarkably, the consumption literature rarely distinguishes consumption for social-status display from sustenance, enjoyment, or other (sometimes overlapping) motivations for consumption. Social status consumption is a zero-sum game, which drives competing individuals or groups toward higher consumption—ending not with "need satisfaction" but only with exhaustion of an individual's resources.

Acting alone, each individual competing for status seeks to make the best of his or her position. But satisfaction of these individual preferences itself alters the situation that faces others seeking to satisfy similar wants. A round of transactions to act out personal wants of this kind therefore leaves each individual with a worse bargain than was reckoned on when the transaction was undertaken, because the sum of such acts does not correspondingly improve the position of all individuals taken together. There is an "adding-up" problem (Hirsch, 1976).

LEVELS OF CONSUMPTION

Next we address relative levels of consumption across societies. We compare consumption of the main two throughputs of environmental interest, mass and energy, and further divide mass throughputs into recycled and nonrecycled categories. Energy use has been thoroughly studied in a number of indigenous societies. Total mass throughputs of indigenous societies have not been studied explicitly, but we can make estimates from existing ethnographic data. The bulk of the mass used by indigenous peoples is biodegradable and recycled by biological processes. Wood, hide, reed or bamboo, foodstuffs, and such, when discarded, degrade and feed biological cycles. These societies also create a one-way (nonrecycled) flow of materials for stone tools, such as chert, flint, and obsidian. Ceramic vessels can survive 10,000 years before reintegration into the soil, so we consider them nonrecycled as well.

Table 4-6 shows four types of societies, with estimates for energy, nonrecycled material, and recycled materials. Note that hunter–gatherers function at two orders of magnitude less energy and one order of magnitude less materials than the United States. Swiddeners (based on Beckerman, 1976) use about the same level of materials as does the United States but differ from the United States in that over 99 percent of the materials are recycled. The supposedly modern concept of "sustainability" has been achieved in most hunting–gathering and swidden agricultural societies, as evidenced by the fact that many of these societies can be shown to have run their systems of materials and energy throughput in the same locations for millennia. These societies modify their

TABLE 4-6 A Rough Quantitative Comparison of Energy and Materials Use Across Diverse Types of Societies

	Energy (kW/capita)	Nonrecycled Materials (kg/capita/day)	Recycled Materials (kg/capita/day)
Hunting and gathering	0.11	0.035	3.6
Swidden horticulture	0.25	0.15	50–100
Agriculture in a developing country	1-3	0.5	4–50
U.S.A.	11	56	2.7

NOTE: The full paper explains how the quantities were calculated or estimated.

environments initially—especially swiddeners—but then continue in the same location for very long periods without continuing environmental degradation. Their long-term durability within the ecosystem is not matched by durability in contact with the global political economic system—upon this contact they are quickly absorbed into the nonsustainable world economy.

HOW CAN ONE COMPARE QUALITY OF LIFE?

Low-consumption societies are not very relevant if the life they live is "nasty, brutish and short" [Hobbes, 1968(1651):186]. Comparing quality of life across societies is fraught with problems, but anthropologists have developed some measures. To summarize briefly, our paper suggests potential candidates such as nutrition, health, life-span, work time vs. leisure time, in- vs. out-migration ("revealed preference" for a given society), and relative perceived quality of life by ethnographers.

Of course, this is not a complete list of all the measures we would like to have. What the above measures have to recommend them is that they are in available data—ethnographic, archaeological, human biological, or paleoarchaeological records. When objective measures are applied to compare the quality of life across widely divergent societies, the results are surprising. We concentrate here on work time and health; other measures are covered in Kempton and Payne (forthcoming).

RESULTS OF COMPARISON

Regarding health measures, studies of skeletal remains show that health declined—not improved as might be expected—after transitions from hunting-gathering to early agriculture, then from early agriculture to archaic states. Decreases in health occurred due to the greatly reduced range of plant species eaten, social stratification resulting in separation of decision makers from the bulk of the population, and high population densities leading to infectious disease (Diamond, 1987). Health did not improve markedly until the nineteenth and twentieth centuries, as a result of public health measures in the cities and the advent of modern antibiotics. Life-span is longer in industrialized societies than in any of the indigenous societies we discuss.

One component of the "most toys" myth is that the devices we consume reduce our work time—i.e., that life is easier today than in earlier historical periods or in technologically primitive societies. Regarding earlier historical periods, Juliet Schor has challenged the myth of less work in the modern era. She takes the comparison back to medieval time (Schor, 1991), finding a large increase in work time during the industrial revolution and a decline back to medieval levels during the twentieth century. We feel that the work-time comparison gets more interesting when extended to indigenous peoples. Several sources demonstrate that sustenance requires less work in primitive societies than in our own.

In one study of indigenous people, Johnson (1978), compares middle-class France with the Machiguenga, a horticultural group in the Peruvian rain forest, another society autonomous from the global economy. The middle-class French had 10 hours of free time per day, and the Amazonian people had 14 hours of free time per day; free time in each case included about 8 hours of sleep. Qualitatively, Johnson made parallel observations about his own time sense while living there: "[In] their communities . . . I sense a definite decrease in time pressure . . . when I return home [to the United States] I am conscious of the pressure and sense of hurry building up to its former level" (1978:53). Other studies of time required for subsistence are reported in Sahlins (1972). Hunting–gathering societies often require only 3-4 hours of work to provide an ample and varied diet. In sum, hunter-gatherers and swidden agriculturalists work less and have more leisure than citizens of industrialized societies. Other studies of contemporary indigenous peoples being drawn into the market economy similarly demonstrate the forces that lead to incorporation into market-based, higher-consumption lifestyles. The attraction at time of entry occurs despite eventual degradation in the incorporated peoples' quality of life (e.g., Barlett and Brown 1985; Bodley, 1990).

To briefly consider urban-industrial society, many authors argue that many of the historical developments of the past century have been inconsistent with a higher quality of life for individuals. Historians like Wiebe (1967), Hughes (1989), and Marcus and Segal (1989) have traced the rise of the technological society during the twentieth century and identified its defining focus to be the growth of large-scale organizational systems. Organizational systems can be defined as operational structures that provide their members with efficient means of achieving given ends [compare the administrative theories of Simon (1957) and the economic theories of Galbraith (1967)].

Social critics such as Mumford (1934, 1967, 1970) and Ellul (1964) have argued that these organizational goals result in isolated, dehumanized individuals, while benefiting the organization itself. Furthermore, organizational theorists have argued that large-scale organizations prevent individuals from developing fully on a psychological level. Drawing heavily on the work of psychologists such as Jung and Marcuse, Denhardt (1981) has argued that there is a fundamental tension between the individual and collective psyches. This presents a problem of integrating the individual and collective psyches into a self-actualized whole. In this view, the development of organizational systems has led to a collective psyche that values certain aspects of the human psyche (rationality, instrumentality, etc.) at the expense of others (emotion, expression, etc). The repression of these emotive values hampers individual development. Because of the structure of organizations, therefore, social and individual development is inhibited.

The perspectives of these authors suggest that organizations have become autonomous actors in our society, furthering their own aims rather than human welfare. When people believe that the efficient production of goods is the means for improving their quality of life, this pursuit makes sense. As people come to recognize the destructive characteristics of the material lifestyle and of the systems that support it, they see organizations as, in many respects, fulfilling organizational aims to the detriment of human individuals. The aspects of life that define us as human—expressive, creative, unique—are those aspects that are in conflict with the needs of organizational structures for efficient operation. It is, finally, organizational operations that are supported by the myth of consumption.

CONCLUSION

Obviously, we do not advocate abandoning fixed communities, agriculture, and modern technology, a change impossible at current world population levels. Rather, we wish to list the following observations from the data outlined here and explored more in the full paper:

(1) Most early societies had consumption levels several orders of magnitude smaller than industrial societies today. However, some indigenous societies had very high levels of per capita materials consumption,

similar in magnitude to the United States today but with virtually all in materials promptly recycled by the biosphere.

(2) By objective indicators other than life-span, the quality of life in some ultra-low-consumption societies seems rather high—higher than the societies they next evolved into, and by many indicators higher than ours today.

(3) Major social transitions can occur if they provide benefits to decision-making elites and greater "fitness" at the societal level (e.g., military advantage or rapid growth and spread of the sociopolitical system).

(4) Increasing the quality of life of the broad masses of individuals is not a criterion by which organizations survive, nor has it been a force determining the direction of social evolution.

ACKNOWLEDGMENTS

We are grateful to Jill Neitzel for major conceptual, literature, and reference suggestions. Steven Beckerman and Thomas Rocek provided important data. Abigail Jahiel, Faith Mitchell, and various participants at the National Research Council workshop provided helpful comments on the argument and logic of drafts of this paper. None of these commentators and contributors are responsible for its content.

REFERENCES

Barlett, P.F., and P.J. Brown
1985/ Agricultural development and the quality of life. *Agriculture and Human Values.*
1994 Pp 175-182 reprinted in A. Podolefsky and P. J. Brown, *Applying Anthropology: An Introductory Reader.* Mountain View, Calif.: Mayfield Publishing.
Beckerman, S.J.
1976 The Cultural Energetics of the Barí of Northern Columbia. Unpublished Ph.D. dissertation, University of New Mexico, Department of Cultural Anthropology.
Bodley, J.H.
1990 *Victims of Progress.* Mountain View, Calif.: Mayfield Publishing.
Denhardt, R.B.
1981 *In the Shadow of Organization.* Lawrence. Kans.: The Regents Press of Kansas.
Diamond, J.
1987 The worst mistake in the history of the human race. *Discover* (May) 64-66.
Ellul, J.
1964/ *The Technological Society.* Translated by J. Wilkinson. New York: Alfred
1954 A. Knopf and Random House.
Galbraith, J.K.
1967 *The New Industrial State.* New York: Signet.
Hirsch, F.
1976 *Social Limits to Growth.* Cambridge, Mass.: Harvard University Press.

Hobbes, T.
    1968/   *Leviathan*. Penguin Classics, 1985 ed. New York: Penguin Books.
    1651
Hughes, T.P.
    1989   *American Genesis: A Century of Invention and Technological Enthusiasm 1870-1970.*
            New York: Penguin Books.
Johnson, A.
    1978   In search of the affluent society. *Human Nature* (Sept):50-59.
Kempton, W., and C. Payne
    in   Cultural and social evolutionary determinants of consumption. In T. Dietz, ed.,
    press   *Environmental Impacts of Consumption.*
Marcus, A.I., and H.P. Segal
    1989   *Technology in America: A Brief History.* San Diego, Calif.: Harcourt Brace
            Jovanovich.
Mumford, L.
    1934   *Technics and Civilization.* San Diego, Calif.: Harvest/Harcourt Brace and
            Company.
    1967   *Technics and Human Development. The Myth of the Machine,* Vol. 1, 1st ed. New
            York: Harcourt Brace Jovanovich.
    1970   *The Pentagon of Power. The Myth of the Machine,* Vol. 2, 1st ed. New York: Harcourt
            Brace Jovanovich.
Sahlins, M.
    1972   *Stone Age Economics.* Chicago: Aldine-Atherton.
Schor, J.B.
    1991   *The Overworked American: The Unexpected Decline of Leisure.* New York: Basic
            Books.
Simon, H.A.
    1957   *Administrative Behavior: A Study of Decision-Making Process in Administrative Orga-
            nization,* 2nd ed. New York: Free Press.
Wiebe, R.H.
    1967   *The Search for Order: 1877-1920.* New York: Hill and Wang.

BIBLIOGRAPHY

British Petroleum Company
    1994   *Statistical Review of World Energy.* London: British Petroleum Company.
Cohen, M.N., and G.J. Armelagos
    1984   *Paleopathology and the Origins of Agriculture.* New York: Academic Press.
Crèvecoeur, J.H. St. John
    1782/   Letter from an American farmer. In A. Kolodny, Among the Indians: The uses
    1993   of captivity. *New York Times Book Review.* (Jan): 1.
The Harwood Group
    1995   *Yearning for Balance: Views of Americans on Consumption, Materialism, and the Envi-
            ronment.* Takoma Park, Md.: Merck Family Fund.
Kempton, W., J.S. Boster, and J.A. Hartley
    1995   *Environmental Values in American Culture.* Cambridge, Mass.: M.I.T Press.
Kolodny, A.
    1993   Among the Indians: The uses of captivity. *New York Times Book Review* (Jan): 1.
Lee, R.B.
    1969   !Kung bushman subsistence: An input-output analysis. In A.P. Vayda, ed., *Envi-
            ronment & Cultural Behavior.* Austin, Tex.: University of Texas Press.

1979   *The !Kung San.* Cambridge, England: Cambridge University Press.
Lightfoot, R.R.
    1994   The Duckfoot site. *Archaeology of the House and Household,* Vol. 2; Cortez, Colo.:
            Crow Canyon Archaeological Center.
Rappaport, R.A.
    1971   The flow of energy in an agricultural society. *Scientific American* 225(3):117-122.
U.S. Bureau of the Census
    1992   *Statistical Abstract of the United States: 1992,* 112th ed. Washington, D.C.: U.S.
            Department of Commerce.
Wernick, I.K., and J.H. Ausubel
    1995   National materials flow and the environment. *Annual Review of Energy and Envi-
            ronment* 20:463-492.

5

Strategies for Setting Research Priorities

Paul C. Stern, Thomas Dietz, Vernon W. Ruttan,
Robert H. Socolow, and James L. Sweeney

The topic "environmental impacts of consumption" suggests a great variety of possible research directions, a few of which are illustrated in Chapters 3 and 4. Because of this variety and the persistent confusion about definitions of consumption (see Chapter 2), it is useful to apply an importance criterion in considering priorities for research on consumption. The highest-priority research questions should be those that concern aspects of consumption with major environmental effects.

This chapter discusses three strategies for setting research priorities so as to direct attention to the most important links between consumption and the environment—that is, to the factors that account for the kinds of consumption that have major deleterious effects on the environment or that may have such effects in the future. It focuses especially on a research strategy that first considers the major environmental effects of consumption and then reasons from environmental effects to human causes. It also discusses two other strategies: one that proceeds from human causes to their environmental effects, and one that begins with possible policy interventions and considers their effects.

ENVIRONMENT-FIRST STRATEGY FOR SETTING RESEARCH PRIORITIES

Human activities that directly perturb key properties of the biophysical environment are important research topics under an importance criterion. Such activities have been referred to as proximate causes of environ-

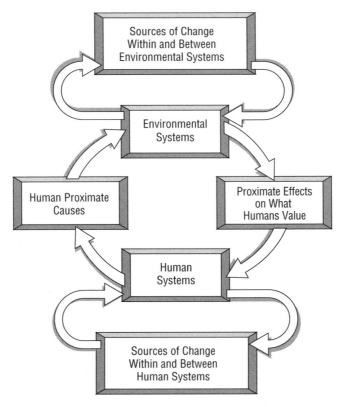

FIGURE 5-1 Interactions between human and environmental systems. SOURCE: National Research Council (1992).

mental change (see Figure 5-1), to distinguish them from other activities, called driving forces, that have environmental effects only through their influence on proximate causes (National Research Council, 1992). Human beings proximately cause environmental change when they burn fuel and thus release combustion products into the atmosphere; when they irrigate land with river water, changing the composition of the water and the height of the water table; when they cut trees or otherwise alter land cover and the habitats of biota; and in countless other ways. The immediate purposes for which they do these things (travel, crop production, construction, etc.) and the deeper reasons behind those purposes (subsistence, financial gain, self-expression, status seeking, etc.) cause environmental change only indirectly.

This line of thinking, from environmental effects to proximate causes to driving forces, suggests a promising and systematic strategy for selecting research that is likely to have relatively large practical importance. The strategy begins by identifying the proximate causes of an environmental condition of concern (the human activities that have the greatest direct effect on it) and then addresses, in a more or less logical order, the following questions: Which actors are responsible for these activities? What driving forces govern their behavior and how do these forces affect each other? What are the trends in these activities and their driving forces over time? How might the activities be changed, if change is desired? The strategy has been described in detail in previous work on the causes of global environmental change (National Research Council, 1992) and is summarized here as it applies to questions of consumption.

Identifying Proximate Causes of Environmental Change

A logical first step in employing this strategy is to identify important environmental properties that human activity may be changing. For example, for climate change, the key environmental properties are atmospheric concentrations of carbon dioxide and other greenhouse gases, and the albedo of the earth. Information on such properties comes from research in the relevant natural sciences, which helps set the agenda for research on human activity by directing attention to those human activities that directly affect the important environmental systems.

The proximate causes of changes in greenhouse gas concentrations are known to a good first approximation. The most important single one by far is the burning of fossil fuels which, by releasing carbon dioxide and methane, is currently responsible for about half of all anthropogenic greenhouse gas emissions. Other important proximate causes include cattle raising, certain land use changes, such as clearing forests and creating or eliminating wetlands, and the various activities that release chlorofluorocarbons. (Some land use changes also affect global climate by altering the earth's albedo.) The uncertainties in estimates of the importance of each of these causes can be reduced by further research on the relevant biophysical processes. Nevertheless, the broad picture can be summarized well enough for the purposes of setting initial priorities for research on human activities. The summary is graphically represented in Figure 5-2, which presents the proximate causes as the limbs of a tree, with their thicknesses indicating their relative contributions to climate change induced by greenhouse gas emissions. Greenhouse gas emissions provide a good illustration of the strategy of reasoning backward from environmental conditions to their causes.

For many other major environmental changes, the relative importance of various human activities as proximate causes is not as well understood. For example, the proximate causes of species extinctions in-

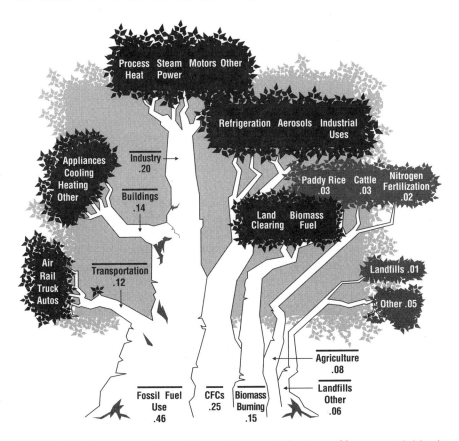

FIGURE 5-2 A representation of the relative contributions of human activities in the late 1980s to global emissions of greenhouse gases. NOTE: CFC = chlorofluorocarbon. SOURCE: National Research Council (1992).

clude a variety of human activities that directly reduce habitat for endangered species. Many scientists believe that clearing of moist tropical forests is the leading cause of species extinction because of the extent of deforestation and because that biome is home to the greatest number of species found nowhere else. But the global rate and distribution of extinctions are still highly uncertain, as is the relative importance of different kinds of land use changes. Still, the analytical strategy of reasoning back from environmental effects to proximate causes remains viable. It will be necessary to revise the research agenda, of course, as knowledge improves about precisely which human activities are the major proximate causes of particular environmental effects.

Estimating the magnitude of the main proximate causes may require difficult conceptual, analytical, or empirical work. To estimate strato-spheric ozone depletion, we need information on the ozone-depleting potential of a fairly large number of chemical compounds and their by-products over periods of time, as well as their rates of release. The most important proximate causes are identified by conducting detailed calcula-tions: they are those activities that directly result in releases with the greatest total ozone-depleting potential for a time horizon of concern. Analysis of the proximate causes of loss of biodiversity is much more difficult. It requires developing acceptable definitions and measures of biodiversity (e.g., should it be defined only as extinctions of species or also in other ways, such as loss of ecological niches or of genetic diversity within species; if the other concepts are important, how can they be mea-sured?). This analysis also requires studies of the effects of particular kinds of human activities on each index of biodiversity. Classifying hu-man activities for this purpose is not straightforward. For example, land use conversion does not have a unitary effect on biodiversity. Its impact depends on the ecosystem in which the land is converted, the new uses to which it is put, and also on such factors as the size, shape, and geographic relationships among unchanged parcels and those that are put to new uses. Because of such complexities, it may be misleading or inappropriate to measure threats to biodiversity in terms of a hypothetical average hect-are of land conversion.

Determining which human activities have major proximate effects on environmental systems can point social and behavioral researchers to the broad types of activity that are likely to be most important in terms of a particular environmental change. For example, once the ozone-depleting potentials and the rates of emission of the various ozone-depleting com-pounds were calculated, it became possible to calculate the overall impact of each gas on ozone depletion, to consider the human activities that release each gas, and thus to identify the most ozone-depleting activities. This kind of analysis is important for choosing which kinds of consump-tion to study. Without this sort of analysis, behavioral researchers have often expended their efforts on studying human activities that are plausi-bly environmentally relevant and that they could easily address with concepts from their disciplines, but that did not contribute greatly to the environmental problem at hand (Stern and Oskamp, 1987).

An alternative procedure for identifying important proximate causes of environmental change is to focus first on major materials and energy transformations and to estimate the environmental impacts of each. This is the approach used by Wernick and Allen in their contributions to Chap-ter 3. This approach is useful for studying the human causes of environ-mental conditions that have been linked to particular transformations of materials or energy. For example, Allen's approach can, in principle, be used to examine any environmental effect that has been attributed to a class of chemical compounds by analyzing flows of the particular compounds most strongly implicated in producing the effect. The difficulty in practice is to identify all the compounds with important direct or indirect effects. Analysis of the human activities that result in production of the responsible chemicals may then identify the most important targets for reducing the environmental effect. For ecological affects, such analysis may need to look not just at the actions that introduce a chemical into the biosphere but more specifically at the actions that introduce it in bioavailable forms. For example, most of the lead that human activity in the United States introduces into the environment is contained in motor vehicle batteries, and the great majority of this is repeatedly recycled. The far smaller amounts of lead in paint, gasoline, and shot have had more serious impacts because of their bioavailability.

Disaggregating the Proximate Causes

When important proximate causes have been identified, the next analytic step is to disaggregate them to reflect the relative contribution of the many actors and activities that are responsible. There may be many sensible ways to disaggregate a particular proximate cause. Fossil fuel burning can be subdivided according to parts of the world (countries or regions), economic sectors (transportation, industrial, commercial, etc.), immediate purposes (space heating, moving people or freight, generating electricity, etc.), types of actor (households, firms, government), or in other ways, depending on the analytic purpose. For important proximate causes, it may be useful to disaggregate in more than one way, creating categories that involve several types of division (e.g., transportation to work, commercial lighting). An impact criterion can help guide decisions about how far to disaggregate. That is, it makes sense to use finer divisions for types of actions or actors associated with major environmental impacts. It follows that the appropriate disaggregation will depend on the society being examined.

Disaggregation can be represented in a tree diagram, as Figure 5-2 shows for greenhouse gas emissions, or in a table. Table 5-1 disaggregates carbon dioxide emissions in the United States by economic sector and purpose within each sector. It shows why, from the standpoint of climate change, much more analytic attention is warranted for understanding the purchase and use of automobiles and light trucks (i.e., personal transportation) than for understanding the purchase and use of water heating systems for buildings.

It can be useful to disaggregate certain activities even further than shown in the table, as a way to direct researchers' attention. For example,

TABLE 5-1 Disaggregation of Carbon Dioxide Emissions by Economic Sector and End Use (percentages, United States, 1987)

End Use	Economic Sector (%)			
	Industrial	Buildings	Transportation	Total (%)
Steam power, motors, appliances	19	7		26
Personal transportation (automobiles, light trucks)			20	20
Space heating	1[a]	16		17
Freight transport (heavy truck, rail, ship, other)			7	7
Heating for industrial processes	6			6
Lighting	1[a]	5		6
Cooling	0[a]	5		5
Air transporting			5	5
Water heating		3		3
Other	5			5
Total	32	36	32	100

NOTE: U.S. data are unrepresentative of world energy use in various ways. However, the United States is responsible for approximately 20 percent of global CO_2 emissions.

[a]Two percent in the single category of heating, ventilating, air conditioning, and lighting was allocated 1 percent each to heating and lighting.

SOURCE: National Research Council (1992).

psychological research on energy use in the late 1970s was mainly addressed to the study of daily behavior in residences, such as turning off lights in unoccupied rooms and altering thermostat settings to save on heating and cooling. An analysis that disaggregated residential energy use according to the energy-saving potential of different household activities showed, however, that infrequent purchase decisions, such as of insulation, furnaces, and major appliances, had significantly greater influence on household energy use than the parallel daily behaviors involving thermostats and appliance use (Stern and Gardner, 1981). The analysis directed behavioral scientists toward the study of a major class of environmentally significant household behavior.

Research to disaggregate the proximate causes has proceeded farthest in the area of energy consumption (the contribution of Schipper in Chapter 3 illustrates some of what can be learned). The strategy of disaggregation can also be applied to other environmentally significant consumption. For instance, in the study of ozone depletion, the use of a particular ozone-depleting gas, such as the chlorofluorocarbon CFC-12, can be disaggregated to reflect its major uses in aerosols, refrigerants, and foams, as well as across countries. If releases of all the ozone-depleting gases were disaggregated in the same way, it would be possible to determine the proportion of total ozone-depleting potential attributable to particular activities such as refrigeration.

Identifying and Analyzing the Driving Forces

The next analytic step is to examine the major classes of environmentally significant action to identify their driving forces. This kind of work is part of the normal province of social and behavioral science, as its aim is to understand the causes of human activity (in this instance, activities that proximately change the environment). These driving forces may be of many types, the main classes of which are human population growth, levels of economic activity, technologies used in actions that affect the environment, political and economic institutions affecting action, and individuals' attitudes and beliefs (National Research Council, 1992). In Chapter 4, Dietz and Rosa examine the first two of these and suggest that the study of what these do not explain may reveal the roles of the other driving forces.

Understanding the driving forces of an environmentally important activity usually presents serious analytical challenges. One reason, as noted in Chapter 1, is that the driving forces act together, often in nonadditive and nonlinear ways. They can affect the proximate causes of environmental change directly and also indirectly by acting on other driving forces. For example, when new chlorofluorocarbon-based technology made air conditioning practical for small commercial and residential uses after 1930, the technological advance set numerous social processes in motion, including large-scale population migrations to hotter climates in the United States, which led to increased use of air conditioning in homes far beyond what would have occurred without the migration. In this way, migration multiplied the effect of technology. In addition, the migration increased fossil fuel consumption for transportation, as the new settlements were less dense geographically than the ones from which people migrated and therefore increased the demand for motorized travel while making mass transit less feasible economically. The migration also shifted the balance of representation in the U.S. Congress in favor of regions more dependent on air conditioning and automobiles, a change that may have made it more difficult to get political support for policies to increase the cost of energy consumption (see National Research Council, 1992:54-60). Thus, the environmental effect of technology was both direct and indirect through such variables as population change and policy. This example illustrates how the driving forces can influence each other: any single factor that seems to explain anthropogenic environmental change is likely to be affected by various other driving forces as well. This is one reason it is difficult to disentangle the causation of environmentally significant consumption.

A second analytic difficulty is that the forces that drive environmental change also respond dynamically to that change. For instance, intensified agricultural production can increase soil erosion or pest infestations, which reduce agricultural productivity until effective changes are made in managing the environmental changes that the agricultural practices created.

A third difficulty is that the relationships among the driving forces generally vary between places and over time. Deforestation has had a different collection of causes in different countries and at different times (Williams, 1990). The long-term environmental effects of a technological development can be very different from the short-term effects. The use of chlorofluorocarbons for refrigeration, for example, was an environmental and public health benefit at first because it replaced ammonia and other toxic or explosive refrigerants. The negative effect on stratospheric ozone became detectable only over a longer time scale and after the level of use had greatly increased.

Although the task of untangling the causality of social phenomena such as those of consumption is difficult, it is a familiar one to social scientists. Experience suggests that the relative importance of the driving forces is likely to be quite situation-specific, even for relatively well-understood activities like those involved in fossil energy use within a single sector of a single economy. For example, the factors determining investment in energy-efficient technologies for the U.S. residential sector are different for occupant-owned and rental housing, for lower- and upper-income households, and probably also for investments in wall insulation and in energy-efficient appliances (e.g., Socolow, 1978; Ruderman et al., 1987; Stern et al., 1986). Research that begins by building relatively narrow areas of knowledge about particular temporal and spatial scales is a first step toward developing a body of contingent generalizations that can make sense of broader bodies of data.

ENVIRONMENT-FIRST STRATEGY AND OTHER APPROACHES

The above discussion and several of the contributions to Chapters 3 and 4 illustrate the value of an environment-first strategy for directing research. Because this strategy begins with important environmental changes and only later ventures into the social phenomena that may provide deeper explanations, following it has a high probability of yielding results that meet a criterion of environmental importance. The environment-first strategy is also valuable because it can help direct social scientists' attention toward human activities and choices that, from an environmental standpoint, are particularly important to understand. As the contributions to Chapters 3 and 4 illustrate, there are two variants of this strategy. One works backward from transformations of materials and energy to their proximate causes in human activity and then to their driving forces. Wernick's analysis (Chapter 3) of materials transformations exemplifies this approach. The other begins with human activities, assesses their importance for particular environmental conditions, and if the activities are sufficiently important environmentally, proceeds to analysis of the proximate causes. The work of Schipper (Chapter 3) on travel and Lutzenhiser (Chapter 4) on home energy use are examples.

Despite the strengths of the environment-first research strategy, other research strategies should also be pursued because they may be able to develop environmentally important knowledge that is not likely to emerge from environment-first research. One useful strategy is to begin with possible policy interventions. This strategy is exemplified by research on the effectiveness of price signals, information, and regulatory constraints for reducing consumer demand for environmentally damaging goods and services or for changing producers' behavior so as to reduce the environmental impact per unit of output. This strategy can have great practical value: for policy purposes, importance is not determined by which human activities do the most environmental damage but by the ones that can yield the greatest environmental improvement in response to practicable interventions. Research on intervention techniques helps determine what is practicable and cost-effective, identifies the possible secondary effects of interventions and the tradeoffs involved, and contributes to general knowledge about how environmentally significant human activities change. Because this strategy is most often employed to evaluate ways to change consumption, most of its applications are outside the scope of this volume, and therefore no examples are included.

Another alternative to looking at environment first, and one that is particularly relevant to the question of what drives environmentally important consumption, is to focus first on social processes and then examine their environmental effects. In Chapter 4, the reports by Wilk and by Kempton and Payne are examples. Wilk asks whether emulation of Western styles of consumption drives the styles adopted by people in developing countries. Although the evidence so far is inconclusive, the question is environmentally important. Kempton and Payne consider the impacts of grand transformations in human history both on the environment and on the quality of human life. In doing this, they raise the question of what might be accomplished by future major social transformations.

There are, of course, innumerable social processes that could be examined for their environmental impacts. The social-process-first research strategy deserves special attention because it is the typical approach social scientists take when they begin to work on environmental problems. Social researchers are likely first to try to understand anthropogenic environmental changes as resulting from the social processes they understand best. As the above examples suggest, this approach can identify some potentially important driving forces of environmental change that the environment-first strategy would be unlikely to uncover. However, the results from research following this strategy should be carefully judged against a criterion of environmental importance.

Experience suggests that although the society-first research strategy can yield fresh insights, it is also likely to produce many findings of limited or uncertain environmental importance. One reason is that the environmental importance of a social phenomenon such as emulation, international trade, economic inequality, or slow transformations of basic social values cannot be estimated until careful analysis has been done. This analysis requires a long time series of multiple variables starting in the past and is likely to reveal that some of the driving forces have only limited environmental effects. Another reason is that it is difficult to evaluate the environmental significance of a single driving force independently of other driving forces. Because of the interactions of the driving forces and their interplay over time, it is inherently difficult to demonstrate the environmental importance of any indirect cause of environmental change, particularly one acting over long time periods.

Generally, it is easier to defend importance claims for driving forces that are linked to environmental effects by relatively short causal chains. For example, Schipper and his colleagues (1989) have argued that increases in female labor force participation in affluent countries increase fossil-energy consumption because they increase the proportion of women holding drivers' licenses, the number of automobiles owned per capita, and the distance traveled. He links changes in household size to residential energy consumption through number of households and per capita volume of heated and cooled residential space. These are strongly plausible claims, but even these require careful analysis to support them. For example, the increased energy demand caused by the need to heat or cool more space is offset (though only to a small extent) by savings resulting from the space being unoccupied during working hours.

Other claims about driving forces involve longer and more debatable causal chains. For instance, it has been claimed that increasing adoption of so-called postmaterialist values (Inglehart, 1990) and proenvironmental worldviews (e.g., Dunlap and Scarce, 1991; Milbrath, 1984) are affecting and will continue to affect human activity in proenvironmental ways. This hypothesis depends on psychological links from values or worldviews to attitudes and from attitudes to behavior, as well as on the presumption that the specific behaviors that may flow from changed values or worldviews will significantly alter the proximate causes of environmental change. Each such link needs to be established to support the hypothesis of environmental importance; the evidence so far suggests that these relationships do exist, but that each of the links is loose (Gardner and Stern, 1996).

Still longer chains of causation are implicit in hypotheses about the environmental importance of various global social changes, such as the globalization of mass media (see Wilk's report in Chapter 4), increasing worldwide urbanization, the emergence of global markets and global communications, democratization, and the resurgence of expressions of cultural identity (see National Research Council, 1992:156-160). Some of these broad social changes may indeed be critically important to long-term environmental trends, but it is difficult to evaluate claims about their importance because the lines of causation are so indirect and, in many cases, because of the long time lags hypothesized between causes and effects. The first step in evaluating the environmental relevance of research into such claims is to make explicit the causal links and the proximate causes that the social process is hypothesized to influence. That done, it is useful to assess the plausibility and evidential base for each link. For example, the emulation of U.S. consumption patterns in developing countries that Wilk discusses, even if actually occurring, is environmentally important only to the extent that particular kinds of consumption that are environmentally destructive (e.g., home air conditioning) are copied.

The environment-first research strategy may converge with the society-first strategy, and such convergence is usually a sign of research progress. For example, Lutzenhiser's data on energy use identify acculturation of immigrants as a factor that affects energy use in a consistent way, even controlling for other factors. This finding suggests a way of improving energy-demand modeling by taking the effects of international migration into account. Similarly, Schipper's findings on female labor force participation and household size suggest some refinements in energy models. Lutzenhiser also finds racial and ethnic differences in consumption, other factors held constant. This finding suggests that a closer look at racial and ethnic subcultures of consumption might increase understanding of the ways attitudes, beliefs, and patterns of living affect the links between affluence and environmental outcomes.

CONCLUSIONS

Research on the environmental impacts of consumption should include a mixture of investigations using environment-first, policy-oriented, and society-first approaches. Because it is anchored to environmental outcome variables of obvious importance, the environment-first strategy is likely to lead to cumulative research that will, over time, continually refine knowledge about the causes of significant anthropogenic environmental changes. This strategy will also increasingly clarify the human choices, actions, and actors that provide important targets for policy intervention, and it will improve the ability of analysts to make projections of anthropogenic inputs into biophysical systems.

The policy-oriented strategy is likely to result in a different sort of cumulative knowledge, leading to increased understanding of the potential for shaping the forces that drive environmental change. It will improve understanding of the conditions under which various kinds and combinations of policy interventions are effective and the extent to which an intervention that works with one kind of behavior or in one social context may be transferable elsewhere. In addition, it is likely to increase basic knowledge about change in individuals and social systems and thus contribute to social science generally as well as to policy analysis.

The society-first strategy is much less likely to generate cumulative research on environmental conditions, but it has compensating advantages. It will occasionally reveal unsuspected but important anthropogenic influences on the environment, especially indirect and long-term ones. Emulation may turn out to be such an influence. Another may be an apparent trend in some wealthy countries toward occupying leisure time by shopping. Yet another, with the potential to alter the environmental impact of human activity, is a trend in the United States for the management of local resources such a parklands and riverbanks to devolve from centralized institutions to local ones—a change that may shift management priorities.

Identifying such indirect or unsuspected influences on environmental quality may help researchers who are following the environment-first strategy to reconsider and improve their conceptual models. Society-first research may also identify possibilities for interventions that long-range planners can use and that may be more socially acceptable than short-term interventions that appear to some people as drastic action to avert a projected emergency that they doubt will ever arise. Changes in the rules governing resource-management institutions could be such an example.

The society-first approach is also likely to lead to insights of great value for constructing scenarios and projections of future human-environment interactions. In addition, it may help interest more social scientists in environmental issues by showing them clear connections between their normal research concerns and an important public policy issue. Involvement by a greater variety of social scientists and collaboration with natural scientists can not only improve research on environment and consumption but also enrich the social science disciplines by bringing them into contact with new sets of phenomena.

REFERENCES

Dunlap, R.E., and R. Scarce
  1991 The polls/poll trends: Environmental problems and protection. *Public Opinion Quarterly* 55:713-734.
Gardner, G.T., and P.C. Stern
  1996 *Environmental Problems and Human Behavior.* Needham Heights, Mass.: Allyn and Bacon.
Inglehart, R.
  1990 *Culture Shift in Advanced Industrial Society.* Princeton, N.J.: Princeton University Press.
Milbrath, L.
  1984 *Environmentalists: Vanguard for a New Society.* Albany: State University of New York Press.
National Research Council
  1992 *Global Environmental Change: Understanding the Human Dimensions.* P.C. Stern, O.R. Young, and D. Druckman, eds. Committee on the Human Dimensions of Global Change. Washington: National Academy Press.
Ruderman, H., M.D. Levine, and J.E. McMahon
  1987 The behavior of the market for energy efficiency in residential appliances including heating and cooling equipment. *The Energy Journal* 8:101-124.
Schipper, L., S. Bartlett, D. Hawk, and E. Vine
  1989 Linking life-styles and energy use: A matter of time? *Annual Review of Energy* 14:273-320.
Socolow, R.H.
  1978 The Twin Rivers program on energy conservation in housing: Highlights and conclusions. *Energy and Buildings* 1:313-324.
Stern, P.C., E. Aronson, J.M. Darley, D.H. Hill, E. Hirst, W. Kempton, and T.J. Wilbanks
  1986 The effectiveness of incentives for residential energy conservation. *Evaluation Review* 10:147-176.
Stern, P.C., and G.T. Gardner
  1981 Psychological research and energy policy. *American Psychologist* 36:329-342.
Stern, P.C., and S. Oskamp
  1987 Managing scarce environmental resources. Pp. 1043-1088 in I. Altman and D. Stokols, eds., *Handbook of Environmental Psychology.* New York: Wiley-Interscience.
Williams, M.
  1990 Forests. Pp. 179-201 in B.L. Turner II, W.C. Clark, R.W. Kates, J.F. Richards, J.T. Mathers, and W.B. Meyer, eds., *The Earth as Transformed by Human Action.* Cambridge, England: Cambridge University Press.

About the Contributors

DAVID T. ALLEN is Beckman Professor of Chemical Engineering at the University of Texas at Austin. Before joining the faculty at the University of Texas, Dr. Allen was professor and chairman of the Chemical Engineering Department at the University of California, Los Angeles. His research has been in the field of environmental reaction engineering, focusing particularly on issues related to air quality and pollution prevention. He is the author of three books and over 80 papers in these areas. He has received a Presidential Young Investigator Award from the National Science Foundation and an award in industrial ecology from the AT&T Foundation. Dr. Allen is active in developing pollution-prevention education materials for engineering curricula. He received his B.S. degree in chemical engineering from Cornell University. His M.S. and Ph.D. degrees in chemical engineering were awarded by the California Institute of Technology. He has held visiting faculty appointments at the California Institute of Technology and the Department of Energy.

THOMAS DIETZ is professor of sociology at George Mason University. He holds a B.G.S. from Kent State University and a Ph.D. in ecology from the University of California, Davis. He has been elected a Fellow of the American Association for the Advancement of Science and is past-president of the Society for Human Ecology. His research interests include the evolutionary dynamics of culture, the use of scientific information in policy processes, and the determinants of environmentally significant behavior. He is coauthor of *The Risk Professionals* and coeditor of the

Handbook for Environmental Planning and *Human Ecology: Crossing Boundaries.*

FAYE DUCHIN is Dean of the School of Humanities and Social Sciences at Rensslaer Polytechnic Institute. In 1985 she succeeded Wassily Leontief as director of the Institute for Economic Analysis at New York University, where for 20 years she studied economic, technological and social change and their implications for the environment in developed and developing economies. Her most recent book, *Changing Lifestyles: the Social Dimension of Structural Economies*, follows upon an in-depth assessment of strategies for sustainable development prepared for the Earth Summit in Rio de Janeiro in 1992. She has been an AT&T Fellow in Industrial Ecology, a fellow of the United Nations University in Tokyo, an officer of the International Society for Ecological Economics, and a founding managing editor of the international journal, *Structural Change and Economic Dynamics*. She received a B.A. degree in psychology from Cornell University and a Ph.D. in computer science from the University of California at Berkeley.

WILLETT KEMPTON is assistant professor in the College of Marine Studies, University of Delaware, senior policy scientist at the university's Center for Energy and Environmental Policy, and assistant professor in the School of Urban Affairs and Public Policy. His scholarly articles cover topics such as the American citizens' understanding of global climate change, international comparisons of citizens' and policymakers' environmental perspectives, beliefs and behavior regarding home energy, energy efficiency policies, and factors that move citizens to environmental action. He is coauthor of *Environmental Values in American Culture.*

LOREN LUTZENHISER is associate professor of sociology, associate research scientist in rural sociology, and a member of the graduate faculty in environmental science and regional planning at Washington State University. His research interests include the social organization of technical systems and the relationship between culture, consumption, and the natural environment. His work has focused on social patterns of energy use in households, the role of behavior in energy systems, issues in energy policy and environmental justice, and institutional barriers to innovation in energy efficiency. Dr. Lutzenhiser is currently engaged in comparative studies of consumption and policy modeling in the U.S. and Europe. He received a Ph.D. degree in sociology from the University of California at Davis.

CHRISTOPHER PAYNE is a senior research associate with the Lawrence Berkeley National Laboratory, a graduate research assistant with the Center for Energy and Environmental Policy at the University of Delaware, and a consultant to the nonprofit Society for Energy Efficiency. His research interests include energy-efficiency investment behavior in the small commercial and industrial sector, environmental identity formation in the workplace and home, the role of organizations in shaping environmental values, the formation of environmentally beneficial federal procurement policies, and the development of implementation of effective public policy. He has a B.A. in physics with a concentration in technology and policy studies from Carleton College and an M.S. in science and technology studies from Rensselaer Potytechnic Institute. He is currently a doctoral candidate at the University of Delaware's College of Urban Affairs and Public Policy.

EUGENE A. ROSA is professor and chair of sociology, professor of environmental science and regional planning, faculty associate in the social and economic sciences research center, and Edward R. Meyer Distinguished Professor of natural resource and environmental policy in the Thomas F. Foley Institute for Public Policy and Public Service at Washington State University. He is currently secretary of Section K (social, economic, and political sciences) of the American Association for the Advancement of Science (AAAS). His research program has focused on environmental topics—particularly energy, technology, and risk issues—with attention to both theoretical and policy concerns. Among his publications are the coedited books: *Public Reactions to Nuclear Power: Are There Critical Masses?*, and *Public Reactions to Nuclear Waste: Citizens' Views of Repository Sitting*. He received his B.S. from the Rochester Institute of Technology and his M.A. and Ph.D. degrees from the Maxwell School of Syracuse University.

VERNON W. RUTTAN is a Regents' Professor in the Departments of Economics and Applied Economics and an adjunct professor in the Hubert H. Humphrey Institute of Public Affairs at the University of Minnesota. From 1961-1963 he served as a staff member of the President's Council of Economic Advisors, and from 1973-1978 he was president of the Agricultural Development Council. His research has been in the fields of agricultural development, resource economics, and research policy. Ruttan is the author of *Agricultural Development: An International Perspective* and *Agriculture, Environment, and Health: Sustainable Development in the 21st Century*, and the coauthor of *Agricultural Research Policy*. His most recent book is *United States Development Assistance Policy: The Domestic Politics of Foreign Economic Aid*. Ruttan has been elected a fellow of the American Academy of Arts and Sciences, the American Association for the Advancement of Science, and to membership in the National Academy of

Sciences. He received his B.A. from Yale University, and his M.A. and Ph.D. from the University of Chicago.

LEE J. SCHIPPER is a staff senior scientist at the Lawrence Berkeley Laboratory, University of California, Berkeley, and coleader of the International Energy Studies group. He is also associated with the Stockholm Environment Institute. He is on leave to the International Energy Agency, Paris, France, until the end of 1997. Dr. Schipper has authored over 75 technical papers on energy economics, energy use, and energy conservation around the world. His is the coauthor of *Energy Efficiency and Human Activity: Past Trends, Future Prospects,* and *Coming in from the Cold: Energy-Wise Housing from Sweden.* He has focused his research on energy use in households, transportation, and indicators of economy-wide energy utilization. His most recent research has been examining the links among transportation, human lifestyles and mobility, emissions, and the environment. Dr. Schipper received his M.A. and Ph.D. degrees in physics from the University of California, Berkeley.

ROBERT H. SOCOLOW is professor of Mechanical and Aerospace Engineering and director of the Center for Energy and Environmental Studies at Princeton University. His research interests include industrial ecology of metals, anthropogenic modifications of the carbon and nitrogen cycles, and energy end-use efficiency in transportation and buildings. He served as chair of the board of the American Council for an Energy-Efficiency Economy (ACEEE) from 1989-1993. In 1995, he was a member on the Fusion Review Panel of the President's Committee of Advisors on Science and Technology (PCAST). Currently, he is a fellow of the American Physical Society and the American Association for the Advancement of Science, and serves on the board of the National Audubon Society. He is the coauthor of *Boundaries of Analysis: An Inquiry into the Tocks Island Dam Controversy; Energy Conservation: Proceedings of the Soviet-American Symposium, Moscow, June 1985;* and *Industrial Ecology and Global Change.* He is also the editor of *Annual Review of Energy and Environment.* He earned his B.A and Ph.D. from Harvard University in physics.

PAUL C. STERN is study director of the Committee on the Human Dimensions of Global Change and the Committee on International Conflict Resolution at the National Research Council, research professor of sociology at George Mason University, and president of the Social and Environmental Research Institute. In his major research area, the human dimensions of environmental problems, he has written numerous scholarly articles, coedited *Energy Use: The Human Dimension* and *Global Environmental Change: Understanding the Human Dimensions,* and coauthored the textbook *Environmental Problems and Human Behavior.* He is a member of the Working Group on Perception and Assessment of Global Environmental Change of the International Human Dimensions Program on Global Environmental Change and a fellow of the American Psychological Association and the American Association for the Advancement of Science. He has also authored a textbook on social science research methods and coedited several books on international conflict issues. He holds a B.A. from Amherst College and M.A. and Ph.D. degrees in psychology from Clark University.

JAMES L. SWEENEY is chair of the Department of Engineering-Economic Systems and Operations Research at Stanford University. His has focused his research on the application of economic methods and mathematical modeling, particularly to natural resources issues, energy economics, environmental economics, competitive analysis, and policy analysis. He periodically serves as a consultant or advisor to Exxon, Atlantic Richfield, Enerco New Zealand, Todd Petroleum, Shell Petroleum, the American Petroleum Institute, Charles River Associates, the U.S. Environmental Protection Agency, and the U.S. Department of Energy. He has served as coeditor of *Resource and Energy Economics* and currently serves on the editorial board of *The Energy Journal.* He was a founding member of the International Association of Energy Economics and has served as its vice-president for publications. He has written several articles for *Econometrica, Journal of Economic Theory, Management Science,* and *Journal of Urban Economics.* He holds a B.S. degree in electrical engineering from Massachusetts Institute of Technology, and received his Ph.D in engineering-economic systems from Stanford University.

IDDO K. WERNICK is a research associate in the Program for the Human Environment at The Rockefeller University. His current research covers long-term patterns of natural resource use and technology development in the United States and the resulting environmental consequences. Specifically, this work concentrates on analyzing the flow of materials in the U.S. economy. He has also investigated environmental causes of mortality and the technical and political context for community risk assessment. He received his B.S. in physics at the University of California at Los Angeles and his Ph.D. in applied physics from Columbia University for experimental work with the free electron laser.

RICHARD R. WILK is an associate professor of anthropology at Indiana University. He has taught at the University of California at Santa Cruz, and New Mexico State University, and served as a rural sociologist with the U. S. Agency for International Development. His research has focused on household organization and decision making, consumer behavior in developing countries, and sustainable development. He has conducted field research on energy use in households in California, on Western consumer goods in West African markets, and on household decision making, migration, development, and consumer culture in the Central American country of Belize. His work on human nature and decision making in different cultures forms the basis for his recently published book *Economies and Cultures: Foundations of Economic Anthropology*. He received his undergraduate degree in anthropology from New York University and his Ph.D. in anthropology from the University of Arizona.